COMPUTATIONAL MODELING OF VISION

OPTICAL ENGINEERING

Series Editor

Brian J. Thompson

Distinguished University Professor
Professor of Optics
Provost Emeritus
University of Rochester
Rochester, New York

1. Electron and Ion Microscopy and Microanalysis: Principles and Applications, *Lawrence E. Murr*
2. Acousto-Optic Signal Processing: Theory and Implementation, *edited by Norman J. Berg and John N. Lee*
3. Electro-Optic and Acousto-Optic Scanning and Deflection, *Milton Gottlieb, Clive L. M. Ireland, and John Martin Ley*
4. Single-Mode Fiber Optics: Principles and Applications, *Luc B. Jeunhomme*
5. Pulse Code Formats for Fiber Optical Data Communication: Basic Principles and Applications, *David J. Morris*
6. Optical Materials: An Introduction to Selection and Application, *Solomon Musikant*
7. Infrared Methods for Gaseous Measurements: Theory and Practice, *edited by Joda Wormhoudt*
8. Laser Beam Scanning: Opto-Mechanical Devices, Systems, and Data Storage Optics, *edited by Gerald F. Marshall*
9. Opto-Mechanical Systems Design, *Paul R. Yoder, Jr.*
10. Optical Fiber Splices and Connectors: Theory and Methods, *Calvin M. Miller with Stephen C. Mettler and Ian A. White*
11. Laser Spectroscopy and Its Applications, *edited by Leon J. Radziemski, Richard W. Solarz, and Jeffrey A. Paisner*
12. Infrared Optoelectronics: Devices and Applications, *William Nunley and J. Scott Bechtel*
13. Integrated Optical Circuits and Components: Design and Applications, *edited by Lynn D. Hutcheson*
14. Handbook of Molecular Lasers, *edited by Peter K. Cheo*

15. Handbook of Optical Fibers and Cables, *Hiroshi Murata*
16. Acousto-Optics, *Adrian Korpel*
17. Procedures in Applied Optics, *John Strong*
18. Handbook of Solid-State Lasers, *edited by Peter K. Cheo*
19. Optical Computing: Digital and Symbolic, *edited by Raymond Arrathoon*
20. Laser Applications in Physical Chemistry, *edited by D. K. Evans*
21. Laser-Induced Plasmas and Applications, *edited by Leon J. Radziemski and David A. Cremers*
22. Infrared Technology Fundamentals, *Irving J. Spiro and Monroe Schlessinger*
23. Single-Mode Fiber Optics: Principles and Applications, Second Edition, Revised and Expanded, *Luc B. Jeunhomme*
24. Image Analysis Applications, *edited by Rangachar Kasturi and Mohan M. Trivedi*
25. Photoconductivity: Art, Science, and Technology, *N. V. Joshi*
26. Principles of Optical Circuit Engineering, *Mark A. Mentzer*
27. Lens Design, *Milton Laikin*
28. Optical Components, Systems, and Measurement Techniques, *Rajpal S. Sirohi and M. P. Kothiyal*
29. Electron and Ion Microscopy and Microanalysis: Principles and Applications, Second Edition, Revised and Expanded, *Lawrence E. Murr*
30. Handbook of Infrared Optical Materials, *edited by Paul Klocek*
31. Optical Scanning, *edited by Gerald F. Marshall*
32. Polymers for Lightwave and Integrated Optics: Technology and Applications, *edited by Lawrence A. Hornak*
33. Electro-Optical Displays, *edited by Mohammad A. Karim*
34. Mathematical Morphology in Image Processing, *edited by Edward R. Dougherty*
35. Opto-Mechanical Systems Design: Second Edition, Revised and Expanded, *Paul R. Yoder, Jr.*
36. Polarized Light: Fundamentals and Applications, *Edward Collett*
37. Rare Earth Doped Fiber Lasers and Amplifiers, *edited by Michel J. F. Digonnet*
38. Speckle Metrology, *edited by Rajpal S. Sirohi*
39. Organic Photoreceptors for Imaging Systems, *Paul M. Borsenberger and David S. Weiss*
40. Photonic Switching and Interconnects, *edited by Abdellatif Marrakchi*
41. Design and Fabrication of Acousto-Optic Devices, *edited by Akis P. Goutzoulis and Dennis R. Pape*
42. Digital Image Processing Methods, *edited by Edward R. Dougherty*
43. Visual Science and Engineering: Models and Applications, *edited by D. H. Kelly*
44. Handbook of Lens Design, *Daniel Malacara and Zacarias Malacara*
45. Photonic Devices and Systems, *edited by Robert G. Hunsperger*

46. Infrared Technology Fundamentals: Second Edition, Revised and Expanded, *edited by Monroe Schlessinger*
47. Spatial Light Modulator Technology: Materials, Devices, and Applications, *edited by Uzi Efron*
48. Lens Design: Second Edition, Revised and Expanded, *Milton Laikin*
49. Thin Films for Optical Systems, *edited by François R. Flory*
50. Tunable Laser Applications, *edited by F. J. Duarte*
51. Acousto-Optic Signal Processing: Theory and Implementation, Second Edition, *edited by Norman J. Berg and John M. Pellegrino*
52. Handbook of Nonlinear Optics, *Richard L. Sutherland*
53. Handbook of Optical Fibers and Cables: Second Edition, *Hiroshi Murata*
54. Optical Storage and Retrieval: Memory, Neural Networks, and Fractals, *edited by Francis T. S. Yu and Suganda Jutamulia*
55. Devices for Optoelectronics, *Wallace B. Leigh*
56. Practical Design and Production of Optical Thin Films, *Ronald R. Willey*
57. Acousto-Optics: Second Edition, *Adrian Korpel*
58. Diffraction Gratings and Applications, *Erwin G. Loewen and Evgeny Popov*
59. Organic Photoreceptors for Xerography, *Paul M. Borsenberger and David S. Weiss*
60. Characterization Techniques and Tabulations for Organic Nonlinear Optical Materials, *edited by Mark Kuzyk and Carl Dirk*
61. Interferogram Analysis for Optical Testing, *Daniel Malacara, Manuel Servín, and Zacarias Malacara*
62. Computational Modeling of Vision: The Role of Combination, *William R. Uttal, Ramakrishna Kakarala, Sriram Dayanand, Thomas Shepherd, Jagadeesh Kalki, Charles F. Lunskis, Jr., and Ning Liu*
63. Microoptics Technology: Fabrication and Applications of Lens Arrays and Devices, *Nicholas F. Borrelli*

Additional Volumes in Preparation

Visual Communications and Image Processing, *Chang Wen Chen and Ya-Qin Zhang*

Optical Methods of Measurement, *Rajpal S. Sirohi and F. S. Chau*

Integrated Optical Circuits and Components: Design and Applications, *edited by Edmund J. Murphy*

COMPUTATIONAL MODELING OF VISION

THE ROLE OF COMBINATION

WILLIAM R. UTTAL
Arizona State University
Tempe, Arizona

RAMAKRISHNA KAKARALA
University of Auckland
Auckland, New Zealand

SRIRAM DAYANAND
Alias/Wavefront
Toronto, Ontario, Canada

THOMAS SHEPHERD
Rainbow Studios, Inc.
Phoenix, Arizona

JAGADEESH KALKI
Microsoft Corporation
Redmond, Washington

CHARLES F. LUNSKIS, JR.
Rock Ridge Systems
Las Vegas, Nevada

NING LIU
Honeywell, Inc.
Phoenix, Arizona

MARCEL DEKKER, INC. NEW YORK • BASEL

ISBN: 0-8247-0242-5

This book is printed on acid-free paper.

Headquarters
Marcel Dekker, Inc.
270 Madison Avenue, New York, NY 10016
tel: 212-696-9000; fax: 212-685-4540

Eastern Hemisphere Distribution
Marcel Dekker AG
Hutgasse 4, Postfach 812, CH-4001 Basel, Switzerland
tel: 44-61-261-8482; fax: 44-61-261-8896

World Wide Web
http://www.dekker.com

The publisher offers discounts on this book when ordered in bulk quantities. For more information, write to Special Sales/Professional Marketing at the headquarters address above.

Current printing (last digit):
10 9 8 7 6 5 4 3 2 1

PRINTED IN THE UNITED STATES OF AMERICA

This book is the collective product of all of us. There were many others, however, who either directly or indirectly made it possible for us to complete this work. We dedicate this book to these important people in our lives.

For
Lara
G3
Mit-Chan
Liu Yuan Zhen and Liu Huang Houren
R. H. Shepherd, M.D.

From the Series Editor

Modeling and simulation have become extremely important tools in science and technology. Indeed they have become an integral part of the scientific method of interaction between observation, theory, and controlled experimentation—modeling theoretical propositions and simulating proposed experiments. In addition, both modeling and simulation have become vital aids in education and training, particularly as we deal with the more and more complex systems that control our lives and the infrastructure of modern society, whether it be air traffic control, global navigation, power plant operating systems, land usage programs in urban and suburban planning, or transportation systems—the list seems endless. In the end, all these systems support the human decision maker.

One of the major sensors that we have is vision. Sophisticated information is presented to the human *observer*. Thus the detailed modeling of vision is essential if we are to maximize the machine–human interface. The details of such a modeling effort are presented in *Computational Modeling of Vision: The Role of Combination*.

Brian J. Thompson

Preface

Coherence, the power of the grand explanation, not isolated proofs and predictions, gives science its strength and cogency. Understanding is the name of the game.

Owen Gingerich

This volume is a report of the contributions of a group of us who have been working together in the Arizona State University's Perception Laboratory since 1988. Earlier parts of our work were reported in a volume entitled *The Swimmer: An Integrated Computational Model of a Perceptual-Motor System* (Uttal, Bradshaw, Dayanand, Lovell, Shepherd, Kakarala, Skifsted, and Tupper, 1992.)

In the years since we finished the manuscript of the first volume describing our SWIMMER project, the work has continued at a rapid pace. Our goal remains the same: to program our model so that it is capable of visually sensing its environment, interpreting that environment within the limits of its own world model, deciding on a course of action, and then behaving in an adaptive manner. As was our original intent, what we have accomplished was not only an engineering tool but also a theory of vision.

In general we have been guided by the same premises on which the original report on the SWIMMER was founded, namely: (1) There is much to be gained by approaching a modeling task from a broader perspective than is usual in this field. Rather than attempt to fine-tune a single-purpose algorithm, we sought to integrate many different procedures and processes into a comprehensive theory of an entire organism. (2) From a purely psychobiological point of view, a model that integrates and combines many different weak and idiosyncratic processes into a powerful and capable outcome is a more realistic expression of the way the real organic perceptual-motor system works.

In recent years, these two premises have become more generally accepted, not only as a psychophysical credo but also as a physiological one. Two especially important books have been published that review the substantial amount of information supporting the notion of a separation of visual functions into distinct modular processes as well as different anatomical regions. The first of these is Zeki's (1993) extraordinary review of the specialization of function in the visual brain. The second, Stein and Meredith's (1993) treatise on the merging of the senses, is not limited to vision alone but deals with the problem of how the different sense modalities interact on both an anatomical and a behavioral level.

There is one grand idea emerging from all of the work in vision science, including ours, these two books in particular, and the efforts of many other scholars, scientists, and engineers in many different fields of biology, psychology, and computer science. That idea is that the brain cum mind is made up of a number of nearly independent channels and specialized centers that deal separately with the different parts of our sensory input.

Our interest in this book is mainly with the visual processes. The decomposable components of an image, as we elaborate in the first chapter, are its attributes, or independently measurable dimensions. The thesis of this book, and of an increasing corpus of modern visual science, is that input information is first analyzed into its component attributes and subsequently synthesized, combined, or (in more modern terminologies) fused or bound into a complete multiattribute perceptual experience. Specifically, the last decade or so has seen an enormous change in the way we believe the organic vision system operates. Vision (with a capital V) now seems to be better described as a collection of "visions" (with a small v).

To make this argument of a collection of "visions," scientists in this field have traditionally used the tools of neurophysiology or neuroanatomy on the one hand, and of psychophysical and perceptual data on the other. It is obvious, however, as one surveys the literature, that in recent years another research tool has evolved that helps to make this argument. That new tool is the computational model, existing as a series of program steps, operators, and algorithms.

Of course, each of these approaches is limited and incomplete. The leap that is being attempted from the microscopic neuron to the macroscopic perception, for example, is fraught with conceptual and fundamental difficulties as well as with technical ones. The case against an extreme reductionist approach is detailed in an earlier book (Uttal, 1998). There it was argued that neuroreductionism is computationally impossible and that molar psychophysics is totally incapable of analyzing the constituent mechanisms that underlie the several attributes. Similarly, computational modeling, it must be appreciated, is also limited in what it can accomplish. It is all too often ignored that excellent descriptions are not adequate reductive explanations. Nevertheless, it is clear that none of these approaches is, a priori, superior to any other. Most of the difficulties of one approach are mirrored in those of the others. The triumvirate of modeling, neuroscience, and psychophysical approaches, when used collectively and in mutual support, does, however, create a powerful synergism and produce heuristics that can strongly suggest if not rigorously prove.

Thus, this book merges several traditional approaches to vision. The neurophysiological, neuroanatomical, and psychophysical literature are reviewed to identify the existing empirical support for the notion of a modular vision system. Then we report the details of our current modular computer model of our "seeing" SWIMMER to elucidate further the plausible and to eliminate the ridiculous.

In a larger sense, however, not only did the neural, psychological, and computational methodologies interact to produce the particular theoretical orientation that characterizes this book, but so also did those of us who contributed to this book. The putative understanding that has emerged from a prolonged immersion in the problem of vision, utilizing and depending on a variety of methodologies, was also enhanced by the discussions and interactions among a variety of people.

The modeling effort carried out in the Perception Laboratory of Arizona State University's College of Engineering and Applied Science was very much a collective effort, with ideas being passed back and forth among all of us. In each case, however, the specific programming work, as well as the writing, was done mainly by a single individual.

We also want to acknowledge our collective gratitude to the Office of Naval Research for their support for this project during the years 1988–1996. The specific documents that funded our most recent activities are known in naval circles as ONR Grants N000 14-88-K-0603 and N00014-91-J-1456.

We also benefited from the intellectual support of others who are not explicit authors of this book. At Arizona State University, Professor Tom Taylor, from the Department of Mathematics, was a source of wisdom and guidance to us, with his crystal-clear analysis of the sometimes-esoteric mathematical foundations of some of our programming efforts in our weekly interdisciplinary Vision Seminar. Professor Tom Foley, from the Department of Computer Science, acted as an intellectual guide for the work on surface reconstruction. Personally, all of us are grateful to Professor Philip Wolfe, chair, and our other colleagues in the Department of Industrial and Management Systems Engineering for their willingness to provide a technical and intellectual home for a body of work that has drifted increasingly far away from the more conventional forms of experimental psychology.

William R. Uttal
Ramakrishna Kakarala
Sriram Dayanand
Thomas Shepherd
Jagadeesh Kalki
Charles F. Lunskis, Jr.
Ning Liu

Contents

1

Introduction: A Point of View

1.1 THE ISSUE

Although there is still enormous mystery surrounding the profound question of how we see, there have been some remarkable breakthroughs in our understanding of the visual process and its underlying mechanisms in recent years. One of the most important has been the emerging consensus that the visual system seems to operate on the basis of semi-independent channels that separately convey information about the various attributes of a visual scene to the visual brain. Data from anatomical, neurophysiological, and psychophysical laboratories have combined to clarify this essential fact about the organization of the visual system. We now know that none of these channels is capable of carrying a full and complete representation of the external visual scene. Rather, each channel carries only a small portion of the full range of dimensional or attribute information that fully characterizes the external real-world scene. Each channel is inadequate in encoding all of the information necessary for the representative reconstruction of the scene as a fully dimensionalized perceptual experience.

In light of this now-well-established fact—namely, that each channel of a system of separate channels conveys only a part of the full range of visual information—the logical necessity for some kind of recombination process immediately emerges. Somewhere—or at some time—in the central visual nervous system, the information communicated along the separate channels must be reunited (either in time or space) to create the singular, unified richness of our experienced visual world. Investigation of possible and plausible mechanisms for this combining or binding process has become an active field of research in recent years in several of the cognate fields concerned with visual perception.

The general theoretical concept enunciated in the preceding paragraphs is of profound importance in the development of theories of visual perception. The essence of this revolutionary concept is that, first, a multidimensional or multiattribute scene is analyzed into incomplete representations of each of those attributes and that, second, these separated attributes are then, in some yet-unknown manner, recombined to produce the complete perceptual experience. Implicit in this fundamentally important and novel idea is an essential aspect of visual information processing that has long been overlooked—the binding, fusion, or combination of separate pieces of information into a unified percept. The concept of *binding* is a truly significant intellectual breakthrough that has the potential to have a major effect on both experimentation and theory in visual science. It suggests both a new paradigm for experimentation—the interaction and combination of multiple dimensions of a stimulus—and, even more significantly, a monumental redirection of our theoretical attention toward the interactive effects of several dimensions rather than a rapidly fading overconcern with the impact of single dimensions.

Indeed, binding may even be interpreted from some points of view as suggesting that perhaps our past theoretical strategies have been seriously misdirected and misguided (if not entirely unavoidable) steps in the march toward understanding vision systems. So many of our models and theories in the past, both in the organic and computer vision research domains, have persistently attempted to refine models describing the processing of each attribute so that the individual process works independently as nearly perfectly as possible.

Although efforts in this direction may have been necessary initial steps, our concentration on single components or attributes may have ignored the real nature of visual information processing—the interaction of multiple stimulus attributes as a perceptual response is constructed. In other words, we may have mistakenly sought perceptual perfection at a level at which it simply did not exist in a real biological system. We may have ignored the fundamental fact that perceptual unification and representational precision arise only after attribute binding at higher levels of neural processing.

We argue that the search for fine-tuned perfection in the action of a single dimension may have been a misdirection. An alternative approach to visual theory, one that would be much more in context with the emerging ideas of multiple and fallible channels, assumes that perceptual veridicality (i.e., correspondence between the stimulus and the percept) is not to be found in the peripheral, low-level, attribute-specific communication mechanisms, but in the outcome of a central binding or combining mechanism that merges information from several attributes of a stimulus scene.

Thus, the main topics with which we grapple in this book are the essential binding and fusion processes that are mainly accountable for high-quality perceptual representation of the external scene. It is very important in this context that the reader not overlook the fact that this "mechanism" may not be a specific physical entity existing at a particular place. It may, quite to the contrary, be a temporal unification in which the activity of widely dispersed regions form a unified perceptual experience.

In this book, we present a discussion of the neurophysiological and psychophysical data that support this idea. We then introduce a comprehensive, integrated, unified, computational model of a vision system. The model emphasizes the philosophy of separate attributes, channels, and centers and the idea that this information must be combined or bound at some level for high-quality, full-blown vision to occur. To the greatest extent possible we wish to track the biological data as we develop our model, but we are not slavishly attached to those data. Vision science still does not know enough to limit our work to that knowledge base alone. Often we had to go beyond current knowledge to speculate how something might be done in the nervous system by developing a computer program. Our model is also, therefore, constrained by contemporary computer science developments, available mathematics, and common sense. We work within the current limits of computational analysis and synthesis, and try to develop a reasonable working system based on both these limits and emerging biological findings.

In short, a major shift in the paradigm of visual research is proposed. In addition to studying empirically the interaction among stimulus attributes, we are also suggesting that theoreticians should increasingly emphasize the search for the powerful reconstructive stages where the separate channels of information are recombined and fused to overcome the gross inadequacies in the representations carried by the separate channels.

A major concern for all of us in this field, therefore, is the direction in which vision theory development should be pushed in future years. Does it make sense to continue to attempt to refine our models of the function of the individual attributes or their channels to a level that is beyond the limits of the natural biological system? To the contrary, might we not better assume fragility, fallibility, and error-prone behavior on the part of these independent natural channels and seek the source of our powerful visual skills in the integrating, unifying, combining, binding mechanisms? If this alternative approach (emphasizing the combination processes) is explicitly chosen by visual scientists (or just implicitly evolves), it substantially redirects where our theoretical energies are directed and what will ultimately be considered to be an adequate and "complete" description of visual perception. Such an emphasis would obviously also have a powerful influence on the degree to which attention is directed towards models of sensory attribute unification, integration, binding, and fusion in the coming years.

To make this point clear, it must be emphasized that most contemporary theories of visual perception are essentially microtheories of relatively narrowly defined portions of the total visual problem. Visual scientists traditionally have developed models of texture, depth, or color perception. Until recently, unified models emphasizing the interaction between attributes have been infrequent. The impetus in most contemporary microtheoretical developments is to refine the model of, for example, texture perception so that it is self-sufficient. This "isolationist" approach is usually motivated by a predisposition to make the model as capable as possible to explain a particular perceptual phenomenon within the confines of its own set of premises and axioms.

This kind of modeling approach, we suggest, tends to produce "superbiological" descriptions—theories that overstate the role of the particular attribute processing mechanism and, therefore, misstate the actual nature of the biological reality. The reality of biological systems, we would argue, is that they are actually much less robust than proponents of the usual microtheoretical model of this genre often suggest. The truth of the matter seems to be that perception reaches its fine representation of the visual scene only in terms of the combination of the outcomes of all of the attribute-encoding mechanisms.

"Superbiological" models of subsystems may be of use to engineers attempting to automate some process, but these models are often incorrect as expressions of what is going on in the organic visual system; i.e., they fail as *valid* theoretical explanations or descriptions of what is happening in perception. Inevitably, a superrefined, "superbiological" model leads to superfluous conclusions that deviate from the biological reality. If understanding of the biological system is the goal, as it is for psychologists and physiologists, the theoretical attempt to pack too much power into a putative explanation of a single attribute processor can thus be misleading to a very significant degree. Furthermore, one might also argue that it likewise can be counterproductive for the computer vision engineer. That is, by forcing the engineer to set ever-higher goals and to produce ever more powerful single-attribute processors, it may divert the development of a more elegant and, in the long run, more effective solution to some problem. In other words, a superior solution to understanding organic vision (or producing successful computer vision programs) may actually be forthcoming from concepts based on the interaction and combination of a variety of less competent operators.

The thesis of this book emerges directly from this new perspective emphasizing combination and binding. In this book, as the individual attributes of surface shape or depth, texture, motion, intensity, shading, and color involved in visual perception are considered, the necessity for abandoning the search for "perfect" or individually sufficient models of any of the visual attributes is stressed. We argue on the basis of specific data and existing models for acceptance of the fundamental concept that information being conveyed in each channel is, at best, incomplete and that our theories, models, and understanding of vision may actually be erroneous to the degree that they adhere to outdated notions of the perceptual potency of the individual channels. We contend that these attribute-specific channels contain only imperfect cues, hints, and outlines of the external reality that must be combined or bound with the information being transmitted along other, equally imperfect channels to produce a complete perceptual experience.

We argue further that it is the role of the central unification and combining mechanisms to make sense of the flawed, partial, and incomplete information in the same way that measurement precision emerges from the statistical averaging process. Admittedly, our discussion of the combining and binding process must necessarily be incomplete: There is a paucity of information about the exact nature of the central fusion processes in biological systems. Indeed, some of the recently proposed mechanisms for binding are quite fanciful and often strain credulity. Nevertheless, some ideas have been forthcoming in recent years that might be harbingers of some of the more valid and plausible theories that are certain to emerge in the future. If we cannot yet identify the biological combination mechanisms themselves, there are some very interesting ideas emerging in the computer science field attacking the problem of "data fusion" that may provide some insights and heuristics into how binding might be accomplished in the organic nervous system. We review a variety of theories of binding in Chapter 3.

In this book stressing combination, is the approach that emphasizes binding over the fine-tuning of the function of single attributes totally novel? Of course not! There were suggestions years ago that individual channels encoded incomplete information. For example, for over a century there has been an appreciation that color information is conveyed along a system that has tripartite properties. It was well known that three signals individually conveying virtually no information about the wavelength of the stimulus, and probably traveling along three different "channels," were combined someplace to produce trichromatic (wavelength-encoding) vision. Regardless of the differences in the particular neural code proposed in the Young–Helmholtz and Hering theories, all appreciated that the full, rich perception of the chromatic attributes of the stimulus emerged only as a result of the combination (or comparison) of the univariant information conveyed in the three channels. From three channels, each of which conveys essentially achromatic vision, arose chromatic vision, but only as a result of combinations and comparisons! This is the same fundamental principle that underlies the more general line of thought advocated here. Now, however, the principle of separation and subsequent combination transcends a single attribute such as color and includes all of the different aspects and dimensions of the visual scene. We believe profound insights have emerged as a result of the renaissance of the general principles of initial channel independence and subsequent recombination.

To make this point clear, let's consider the model of trichromatic vision in a bit greater detail. It has been also well known for centuries that color vision would be terribly flawed if one or more of these "channels" was functioning improperly. The *trivariant* theory of vision that emerged from the phenomena of color mixing, color blindness, and "neutral points" stimulated some important, related ideas. Among the most important, of these was the enunciation of the principle of *univariance* (Rushton, 1972). This principle asserted that, because of the breadth of the absorption spectrum of the photochemical contained in a photoreceptor, the information encoded in the stream of neural signals from a single type of photoreceptor (or single neural channel emanating from that type) could encode only the *amount* of light absorbed by the receptor. There is no way that information about the specific wavelength of the light could be represented by a single photoreceptor that had anything other than a monochromatic absorption curve. Even then, each wavelength would have to be encoded by an inefficient system consisting of as many different receptor types as there were discriminably different wavelengths.

However, wavelength is encoded and color is seen with extreme precision with only a few receptor types. This is accomplished by a relational process in which the outputs of multiple photoreceptors (three were shown to be sufficient) with broad—in contrast to narrow—spectral absorption sensitivities were compared with each other. Whereas the individual receptors or channels were essentially "color-blind" (or, more correctly, "wavelength indiscriminate"), the collective interpretation of their combined output was sufficient to represent wavelength and to provide sufficient information for the precise representation of color.

The notion of channel univariance that emerged in color theory can easily be appreciated to be harbinger of a new appreciation of multiattribute and independent channels for texture, intensity, motion, form, and depth (among the most obvious candidate attributes). It is also probably the case that the absence of any one of these other channels degrades the visual image in a way that can be considered to be analogous to "color blindness." Without intensity differences, boundaries and shapes may be indescribable and indiscriminable; without binocular disparity, stereoscopic three-dimensionality may be missing from the percept; and without texture, images may fade quickly from our visual experiences in their entirety. To the well-known "color blindness," therefore, must be added new notions of "texture blindness," "stereo blindness,"[1] or even "form blindness." To the idea that a trivariant code for wavelength can be created from univariant sensors must be added the idea that full, rich, multivariate scene perception can emerge from the combination of sensor processes that, by themselves, would be selectively "blind" to other attributes.

Why have we not appreciated this previously? A major reason is simply that the predominant scientific approach was unidimensional. That is, there was much work to be done just understanding the simplest kinds of psychophysical relationships. It is only in recent years that our science has evolved to the point that interaction, combination, and such esoteric concepts as data fusion and binding could even become a part of the zeitgeist. Another probable reason is that we tended to organize our science into isolated compartments; our taxonomies of visual phenomena were just not correct or fruitful. Just as universities have traditionally organized themselves in departments separated by nearly insurmountable intellectual walls, visual scientists concentrated on understanding color vision or form perception or depth perception in isolation from the other attributes. One only has to peruse the main body of the specific microtheories discussed in the literature to appreciate how theoretical developments followed the constraints imposed by these restricted and rigid conceptual taxonomies or the limits imposed by executable experimental designs.

We suggest that our science is now evolving to a new, multidimensional level of analysis. As we see later, there are new theoretical models that do combine and new experimental designs that do study interattribute interactions. Furthermore, there has also been a change in the scientific culture. Interdisciplinary science is on the upswing, and vision scientists are now more willing to cross the frontiers dividing study of the various attributes than they had been previously.

We must not be too optimistic about this movement toward intellectual synthesis, however; remember that much of this research still goes on in an organization that on occasion cloaks itself in medieval monks' robes and carries weapons of medieval chivalry in their processions of honor. Similarly, many—if not most—vision scientists are still immersed in attempts to tune the theories more finely and to measure the properties of individual attribute operators. Some of this is not only desirable but necessary. However, the action at the furtherst frontiers of knowledge is, in the opinion of some of us, elsewhere.

Computer scientists have a favorite expression, "GIGO" which means "garbage in, garbage out." The idea is that computers cannot take incorrect input information and transform it into correct output information. If misinformation is introduced to the computer, then at the least, misinformation emerges, and at the worst, total chaos. It may be that like so many other analogies between computers and human brain processes, this hoary adage is also dysfunctional. We contend, quite to the contrary, that "garbage"—in the sense of incomplete, irregular, or indiscriminate information—may be very useful to a system that is capable of integrating, combining, or fusing such a batch of poor information. By doing so, we believe, such a system can produce highly accurate and valid perceptual representations of the visual scene. This approach is in the tradition of statistical analysis, so long a mainstay of the social and behavioral sciences, where variability and multidimensional situations are frequent. Indeed, statistical combination is also the basis of many computer models of vision, among which must be included the one emerging from our laboratory (Lovell et al., 1992).

Perhaps in the bright light of this new approach stressing combination, it would not be inappropriate to add another word to the lexicon of computer and biological visual theory. That word would be "GIBO." "GIBO" reminds us that sometimes you can indeed make a "silk purse out of a sow's ear," or, in other words, take "garbage in" and get "beauty out."

Ramachandran (1985) says it in another way. He points out that the visual system acts more like a bag of virtually independent tricks that must be combined to produce the full visual experience. Although his bag of trick is not aimed at the attributes specifically, but rather at solutions to a number of different perceptual illusions, the idea of independent faculties or operators is also implicit in his concept.

The concepts "GIBO" and "bag of tricks" represent another way of characterizing and emphasizing the dominant theme of this book. Although it is necessary to examine briefly some theories that have been developed to describe processing of the individual attributes, a main goal of this book is to suggest how information from those individual-attribute channels might be bound into a unified visual process. In order to do that we must make clear what some of the specific terminology used means. That is the purpose of the next sections of this chapter.

1.1.1 The Nature of Attributes

In the introductory section of this chapter, we set forth a basic organizational principle of the visual system—the unification of quasi-independent channels—that organizes and guides the remainder of this volume. In our previous books, we have found that in so setting the stage for the rest of the book, we sometimes introduce new, unfamiliar words (or new uses for familiar words) or (worse) words used in unconventional ways. The best way to overcome such difficulties is to present a minilexicon at the outset, to make clear the meaning of the key terms used in subsequent discussions. That is the purpose of this section. In the following paragraphs, we consider, in turn, what is meant by *attributes*, *channels*, and *combining*.

The terms *dimensions*, *attributes*, and *cues* are all used in this book to denote the same concept. These terms refer to *separable properties* of the stimulus. It is very important to appreciate the distinction we make between the attributes of the stimulus (such as disparity, wavelength, and luminance) and the dimensions of the perceptual experience, such as depth, hue, and brightness. For each of the stimulus attributes, there is (to at least a first approximation) a corresponding experience, but the experience is often not directly correlated with the attribute. The fascinating geometric illusions that fill our elementary textbooks are examples of the manifold misinterpretations of ambiguous stimuli to which humans are subject. They are also compelling evidence that the correlation between sensory stimuli and perceptual experience is not always high.

Similarly, the terminology that describes the properties of a psychological experience *is* distinct and different from the one describing physical stimuli, even if we often misuse and/or confuse one for the other. There is nothing red, for example, about 700-nm wavelengths of light until it is captured, transformed, fused, and interpreted by an organic visual system.

It must also be kept in mind that there is something more fundamental in this context than simply confusions about the distinctions between vocabulary items. That more fundamental issue is that the main purpose of the visual system, as of any other sensory system, is to provide as *veridical* a representation of the external world in the perceptual experience as is possible. However they may be encoded by neural mechanisms and representations, however different the vocabulary of the scientist and of the perceiving layperson, senses have evolved to communicate information about the "outside" world to the "inside" one in a way that will allow the organism to survive! Thus, the degree to which the sensory systems are successful in their evolved task is the degree to which the dimensions of the percept correspond to the dimensions of the stimulus. (For a fuller elaboration of these ideas, see Uttal, 1973.)

It is also often overlooked that the perceptual dimension most frequently linked to a particular stimulus attribute need not be solely or uniquely determined by that attribute. Thus, for example, subjective hue, saturation, and brightness, although determined predominantly by the attributes of physical wavelength, spectral purity, and stimulus intensity, respectively, are influenced in secondary ways by the other two attributes. Furthermore, the experiences of hue, saturation, and brightness, can be powerfully influenced by contrast with the context of the surround (Land, 1977) and even by high-level cognitive processes denoted by such mysterious processes as "set," "expectations," and "Einstellung." A more modern term for these top-down influences is *cognitive penetration*—a powerful perceptual force whose importance is sometimes minimized in contemporary theories describing simple peripheral explanations and single sensory dimensions. A basic property of vision, often overlooked in simple psychophysical experiments, therefore, is that perception can be heavily modified and modulated by previous experience and memory. Thus, a major principle of visual perception is that there is no necessary one-to-one correspondence between the attributes or dimensions of the physical stimulus and those of the perceptual dimensions.

We must also accept the principle that there are several intervening levels of encoding between the transduction process and the central interpretive parts of the brain. These neural conduction codes provide ample additional opportunity for dissociations and discrepancies between the physical stimulus and the perceptual response. As noted, the abundance of perceptual illusions of color, brightness, movement, and position bear witness to the frequent differences that can exist between a stimulus and the perceptual response.

In spite of these caveats, most contemporary theories of vision are still essentially microtheories of one or another of these attributes. The definition and classification of these attributes, linked as they are to physical units, dimensions, and measuring instruments, is itself a challenging task. Indeed, it may be that there is no alternative system of classification, other than the parameters of the external physical world, that could be used to organize any study of the sensory world. The anchor provided by physical measures, terms, and units gives the psychophysics of perceptual experience a reference that is incomparable to those available to any other area of psychology (cognition, emotion, intelligence, etc.). Unanchored as those other areas are, they can wander off into post hoc storytelling that often bears little resemblance to serious scientific enterprise.

Now let us be more specific and consider exactly what are the attributes of both the stimuli and the visual experience that must be combined to produce a fully dimensionalized visual experience. The following list is reasonably exhaustive.

Intensity and Brightness

The stimulus intensity of a light source refers to the strength, magnitude, or amount of the stimulus, independent of its qualitative nature or the time or place at which it occurs. At the most basic "physical" level, stimulus intensity is measured (with a radiometer) in units of the cumulative wattage of the relevant collection of quanta. Such a *radiometric* measurement is made without regard to any differential sensitivity to wavelength other than that of the radiometer itself. However, if light is being used as a stimulus in a perceptual experiment, it is usually corrected by weighting it with the particular wavelength sensitivity of the receptor being stimulated. In particular, when used in human vision experiments, the radiometric intensity is usually multiplied by a set of coefficients corresponding to the shape of the photopic (light-adapted, cone-dominated) sensitivity curve to give the *visually effective* light intensity. This corrected *photometric* (as opposed to radiometric) stimulus intensity curve calibrates the photic stimulus intensity in terms of the physiologically effective light. That is, it is a measure of the light that can actually be absorbed by the photopigments of the receptors. Stimuli that fall outside of the photopic curve have no effective photometric intensity, even though their radiometric intensity may be quite high. Stimuli falling within the limits of the visible spectrum have effective weights other than zero, but of varying amplitude depending on the wavelength sensitivity of the photopigment at each wavelength.

The particular unit of stimulus radiometric or photometric intensity utilized depends on the geometry of the situation. The method for converting from the radiometric system to the photometric system is straightforward and is described in detail in Uttal (1973).

The intensity (as well as the chromatic pattern) of a two-dimensional image an be represented in a computer memory by reserving one multibit memory ocation (pixel) for each resolvable point on the image. In this regard, it is nportant to point out that even though the image on a CRT display, as pereived by the observer, appears to be continuous, all real imaging systems, ectronic or organic, are actually discrete in nature. In the organic eye, the esolution is much finer than in a computer image because of the very small size of the photoreceptors, but the principle is the same. That is, to be completely accurate, both computer and organic codes for images must be represented by discrete mathematical functions rather than the idealized mathematics of continuous functions. The fact that digital computers are also discrete calculating engines, therefore, makes the computational vision model into a reasonable theory of the human vision system. It is often because the computer's resolution is usually too low (a necessary condition to avoid a computational explosion) that a discrepancy between a computer model and human vision may exist.

The physical intensity of a stimulus, as corrected by the weighted photometric values, is closely associated with the brightness, or subjective magnitude, of the perceptual experience. However, as noted, the association is loose and a wide variety of other factors than stimulus intensity can change the apparent brightness of a stimulus or the time or place of occurrence. The term *brightness* refers to the subjective magnitude of the experience, regardless of the chromaticity of the stimulus. A closely related term—*lightness*—refers to the degree of grayness, along a scale that varies from blackness to whiteness. But these two terms are more or less independent: A light gray can be low on the brightness scale and a dark gray can be relatively bright.

Wavelength and Purity—Chromaticity and Saturation

Chromaticity is a more precise term than the popular, but ambiguous, term *color*. It has two perceptual dimensions (three, if brightness, or subjective amplitude, is included): hue and saturation. *Hue* refers to the perceived dominant "color" (e.g., red, green, or blue) and *saturation* refers to the degree to which this dominant color has been "washed out" and pales to a white (e.g., the sequence red, pink, white represents a progressive decrease in saturation, even though the hue remains constant.) The physical dimensions that most closely relate to chromaticity and saturation are wavelength and purity, respectively.

The hue and saturation space of perceived colors is quantitatively represented by the two-dimensional CIE chromaticity diagram shown in Fig. 1.1. Briefly, the CIE diagram plots the proportion of two of the three component colors on the horizontal and vertical axes (the proportion of the third component is uniquely specified, since all three proportions must add to 1). The boundary of

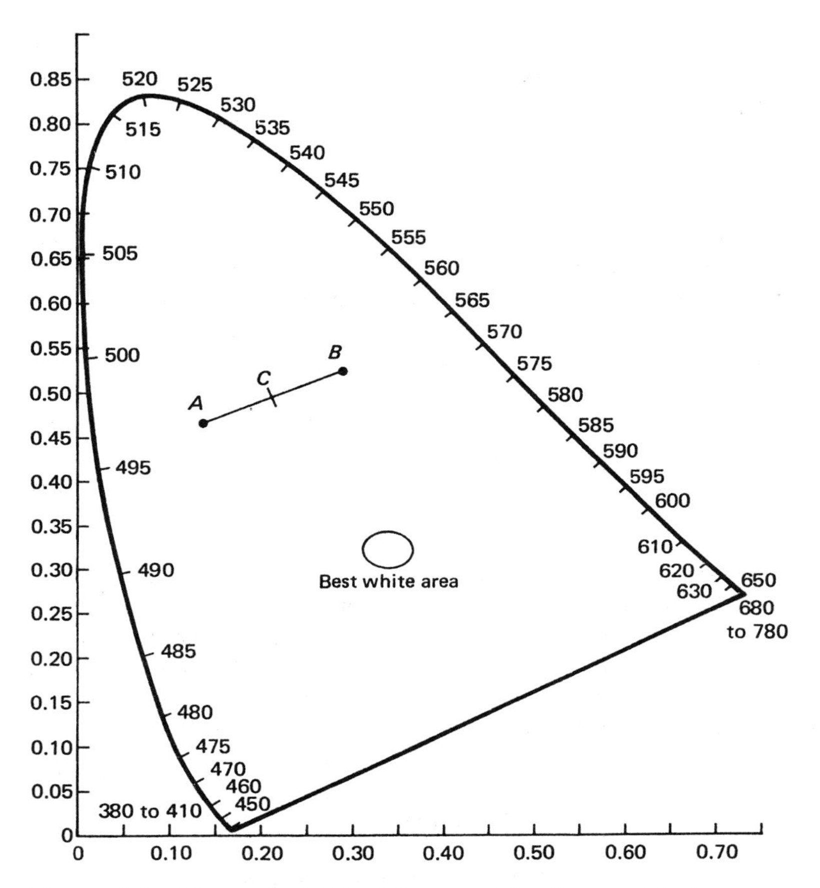

Figure 1.1 The standard 1931 CIE chromaticity diagram. *A* is the proportion of the long-wavelength component of a three-color mixture; *B* is the proportion of the medium-wavelength component; and since $A + B + C = 1$, the proportion of the third component is specified. (From Uttal, 1973.)

the chromaticity space is the locus of the maximally saturated spectral colors, and each point inside the space represents a color experience that is desaturated to some degree.

The correspondence between the physical parameters of the stimulus and the perceptual parameters of the chromatic experience are presented in Table 1.1. Both the primary and the secondary influences of dominant wavelength, purity, and stimulus intensity on hue, saturation, and brightness, respectively, are compared.

Table 1.1 The Relations Between Stimulus and Response Dimensions

	Stimulus Dimension		
Response Dimension	Dominant wavelength	Purity	Luminance
Hue	primary	secondary	secondary
Saturation	secondary	primary	secondary
Brightness	secondary	secondary	primary

In a more practical vein, as noted previously, the storage of the RGB representation of a full-color image requires three times the storage capacity in a computer (and three times the channel capacity in a communication system) than does an achromatic image. Each data storage location for a colored pixel, therefore, must be three times as deep as that required for a monochrome image. That is, in a typical computer system this would require the storage element for a "colored" pixel to be 24 bits "deep" instead of the 8 bits required for a single achromatic pixel. The 8 bits required for a black-and-white image permits 256 different codable (if not perceivable) gray levels. The 24-bit colored image permits 16,777,216 different codable colors, most of which could not be discriminated by the human eye from their immediate neighbors.

Texture

The intensity and wavelength present in an image are meaningful concepts that can be defined at a local point. However, texture—another important attribute of a surface—cannot be defined locally. Texture is a global dimension that becomes meaningful only in terms of the spatial relationships between and among different locations within the stimulus scene. Thus, although the mean intensity or even color values of two texturally different regions may be the same, they may have considerably different textures and perceived qualities as a result of those differences.

Textures, in this context, therefore, are patterns of spatial relationships. Specifically, texture may be defined as the spatial arrangement of the components of a scene. More formally, Texture (T) may be defined as:

$$T = f(I_{x},y,\ \lambda_{x},y)$$

where I_{x},y is the set of intensities at all x,y locations and λ_{x},y is the set of wavelength spectra at all x,y locations. It is often the case that there is some kind of regular order in a textured scene, but regularity is not a necessary property. One textured region characterized by one kind of random organization can readily be distinguished from another characterized by some other kind.

The components from which the textured field may be constructed can themselves be further characterized by the arrangements of local units that could influence the perception of the overall texture. Texture may also be considered to be quantified by the higher- (than first-) order statistics of these smaller regions, following the nomenclature of Julesz (1981). The structured component units of a textured pattern were designated *Textons* by him.

Because of the nonlocal nature of both the individual Texton and the overall pattern of Textons (which we collectively call texture), textural analysis is a much more complicated procedure than is the definition of the chromatic or intensive nature of an individual point in a scene or of a pixel in its computer-stored representation. It is a truism in this field that there is no universal texture analyzer; each such operator is nearly always idiosyncratic and specialized. A texture analyzer that has great success with one kind of texture may fail completely when confronted with another. It is for this reason that the combining or binding of texture analyzers into a system of cooperating units (e.g., see Lovell et al., 1992) has proven to be the best strategy for texture analysis.

Three-Dimensional Shape and Depth

One of the major areas of research interest in recent years has been the recovery of three-dimensional shape information from what are essentially two-dimensional images. Although the most compelling perception of depth comes from a comparison of the slightly different or disparate images impinging on the two retinas (or being captured by two slightly displaced cameras), there are many other cues that can help to define the three-dimensional shape for either a person or a computer. Among the cues of most interest now for recovering three-dimensional shape are the variations in reflected light (shading), motion parallax, perspective, contours produced by the perspective transformation of regular light patterns, and interposition. Shape, of course, is also a nonlocal property of objects. An object may be extended in space, and its shape must be defined by an analysis of the relative positions of all (or a sample of reasonable size and distribution) of the local points on that object.

In more formal terms, three-dimensional shape (S) may be defined by the following expression:

$$S = f(x,y,z)$$

Depth (D), the distance from a point on the shape to the observer (or some other reference point), is simply defined as:

$$D = z$$

for a coordinate system in which the z-axis is perpendicular to the fronto-parallel plane of the observer's face.

The surface shape of an object can, therefore, be specified by a representation of the depth z at each x,y location on the object. The problem is to go from cues such as relative brightness, perspective, and disparity for that shape to specific numerical measures. This is not always directly possible without the specification of additional constraints. In such a situation the problem is said to be ill-posed. An ill-posed problem is not just difficult to solve; it is, in principle, impossible to solve. The mathematics of regularizing ill-posed problems, adding constraints, or bringing in other cues to provide the necessary additional information for solution is currently an active field of research in computer modeling of vision. For a full discussion of this matter, see Poggio and Girosi, 1990.

Some simple surface shapes (e.g., the quadratic and cubic surfaces) can be represented by standard polynomial or parametric equations; other, more complex surfaces must be represented by exhaustive tabular listing of the depths of all of the points on the surface. Complex interpolation programs in which sampled surfaces are connected by curves dependent on the surface normals or local derivatives can sometimes be used to reconstruct the surface piecemeal in a way that allows the surface to be pictorially presented even if a formal expression is not possible.

In the final analysis, the determination of either local depth or global shape is dependent on differences either between images or within an image. Stereoscopic depth analysis attacks the problem by comparing the disparity between two images (captured by two cameras or two eyes) that have been displaced from each other. We describe this process in great detail in Chapter 6 of this book. The recovery of three-dimensional shape (which may be thought of as a pattern of depths) using any other technique is equally dependent on comparisons made between other cues, such as differences in the curvature of a line projected on the object, of the pattern of light reflected from it; or even of different positions it or the observer may have at different times.

Motion

Motion is also an important attribute of a visual stimulus that can help to define its shape. As with any of the other cues, it requires that comparisons be made, but in this case between the image at different times. In fact, extracting depth or shape information from motion can be accomplished with very much the same computational apparatus used for determining stereoscopic depth or range. The major difference is that instead of comparing two images taken at the same instant of time from two different positions, determining shape from motion cues depends on the comparison of two images taken at different times. The exact number of time samples required depends on the viewing situation. However, in the development of computational algorithms, the two processes may be considered to be virtually identical in terms of the steps that are executed to solve their respective problems.

Motion parallax is a generic term for the effect of motion on vision, whether it is the observed object or the observer that moves. In ether case the differences in point of view at different times are the necessary cues for reconstructing the third dimension. A full discussion of shape-from-motion can be found in Ullman (1979) and Watanabe (1998). This book does not deal extensively with motion as a cue for three-dimensional shape.

1.1.2 The Nature of Channels

Throughout this book the term *channels* is frequently used. Indeed, this term, from some points of view, instantiates the central concept of this book. It is, therefore, critical that our readers understand, to as exact a degree as possible, what we mean by a "channel." At the outset, it must be appreciated that the term is used by different authors in different ways and that there may at present actually be no universally accepted definition. Nevertheless, we now focus in on a more precise meaning of the term than has been presented previously.

In its most general neuroanatomical sense, a *channel* refers to a separate pathway from a peripheral receptor to the central nervous system. The term implies the routing of portions of the stimulus information along anatomically or functionally separable communication routes. The anatomic separability of the channels for different attributes may last for only a short distance. For example, the three "channels" for chromatic information in the trichromatic organic eye are structurally separable only within the first layer of the retina—the receptor cones themselves. Beyond the complex synaptic interconnections of the outer plexiform layer, the initial three channels (defined by the distinctive absorption curves of the three cone chemicals—cyanolabe, chlorolabe, and erythrolabe) are transformed into other codes and into new kinds of channels that may carry the same information, but in different ways. Specifically, at the next stage of the vertebrate visual system, the trichromatic information in the receptors is re-encoded into an opponent code in the bipolar cells. The bipolar cell code is subsequently re-encoded in the lateral geniculate body of the thalamus and probably at every other level in the ascending pathway. The complexity of coded representations obviously increases as one probes with microelectrodes at progressively higher levels.

What a "channel" is at these higher levels becomes increasingly difficult to define precisely anatomically because many different cells with many different kinds of sensitivities become involved. The situation may be even more complicated by the fact that many of the neurons carrying coded information about the chromaticity or intensity of a stimulus may also be carrying information about other dimensions or attributes of the stimulus as a result of neural convergence. Channels may, therefore, overlap or may be structurally or functionally redundant. As we see later in this chapter, there is some structural segregation of the channels for the different attributes. However, it is important to appreciate that a "channel" conveying information about a particular attribute may have to be defined in terms of its functional distinctiveness as an addition or alternative to its anatomical segregation.

A key idea to remember in this context is that in spite of the different codes and in spite of the anatomical ambiguity, a channel (i.e., the medium) may do a very good job of maintaining the information representing the attribute (i.e., the message) it has evolved to carry. Thus, even though a color channel may change from a trichromatic to an opponent code, and even though it may become intermingled with some other attribute's "channel," it maintains the essential trivariant nature of the color coding process *and* still represents the spectral properties of the stimulus. The ultimate proof of this assertion is that the psychophysics of human color perception reflects the trivariant nature of the three cones, many neural units distant from the regions in which color is presumably interpreted by the brain—the loci of the psychoneural equivalent.

Similarly, the various coding schemes used in color television channels may change from one scheme to another (e.g., RGB to NTSC to PAL or vice versa) in a way that maintains the meaning (trivariance and appropriate color as well as the intensity and brightness information) of the original scene in each of the coding schemes. Thus, it matters only internally to the "nervous system" of the television display what the language or code is. One code is as good as another in the path to the final display of the scene, as long as the decoding rules are known. Indeed, there is nothing in the picture itself that tells the observer what color coding scheme was used internally. The television code used in the electronic channel, like that of the organic color channel, is invisible to the receiver at the end of the line. However, should the attribute specific channel be severed, then an appropriate kind of selective "blindness" occurs. The invisibility of internal codes to psychophysical probes is an important general principle of perceptual science.

1.1.3 Binding

In this section we present a discussion that we hope leads to an understanding, if not a precise definition, of the concept of binding. Given that there are separate channels for the different attributes, the question now faced is: How is this information recombined or integrated to produce the perceptual experience of a unified scene? Because the concept of binding is so new, because how it is actually implemented is so uncertain, and because it is still controversial whether it is done at all (Caelli et al., 1993; Caelli and Reye, 1993; Grossberg and Mingolla, 1993), a precise and completely satisfactory definition of binding will probably remian elusive for some time.

Among the most important issues on the subject of binding, and thus in understanding visual perception, is whether the combination occurs spatially or temporally. That is, there are two possible ways in which information could be combined. First, it could be that the separate channels of information have to come together at some spatial location in the visual system. This is the easiest kind of binding to conceptualize. All of the separate images, some conveying color, some conveying intensity, some conveying depth, and so on, converge on a region of the brain that merges them into the fully dimensionalized visual experience. Such a spatial integration can easily be modeled and described. Unfortunately, there is no evidence of the total convergence of all image-attribute information at a particular place in the brain. As we discuss in Chapter 2, there is ample evidence of a multiplicity of visual regions that, though interconnected, go about their own business in a more or less independent manner.

There is, of course, a viable theoretical alternative. That alternative is not spatial but, rather, temporal. Rather than seeking a point in space where the channels carrying the coded information concerning the various attributes converges, it is possible to consider the locus of convergence to be a point in *time*. In other words, binding may occur because of synchronous responses rather than propinquitous ones. Given that there is this massive degree of uncertainty, it is possible only to model the process mathematically at the present time. Nevertheless, it should be intuitively clear that what we mean by binding is the combination and integration of the information conveyed along the separate channels into a unified and complete representation of the external scene sensed by an observer. If the observer is a computer and our goal is to develop such a binding mechanism, we have unlimited opportunities for inventiveness. If, on the other hand, our goal is to understand and explain the binding that goes on in an existing organic system, then the task is quite a bit more constrained.

1.1.4 Reduction, Explanation and Description

The use of the terms *reductionism*, *explanation*, and *description* must also be clarified before we can proceed with this discussion. Such definitions have a rich history in philosophy and psychology and certainly have been subject to many different interpretations. Cummins (1983), Dretske (1988), and Haugeland (1981) are among a growing cohort of philosophers who have dealt with the meaning of these words.

Fully appreciating the frailty of any attempt to define terms like these, we nevertheless propose that the following definitions be understood in this section. By *reductive explanation*, we mean a precise, unique, and (hopefully) valid statement of the particular internal mechanisms (neural or cognitive) by means of which a system carries out its function.

Reduction is a more general concept that refers to a kind of theorizing or experimental analysis in which the terms and entities of one level of discourse are interpreted in the terms and entities of a more microscopic one. We understand much of biology because of a reduction to chemistry and much of chemistry because of a reduction to physics. The hope is that we can explain much of psychology by reduction to neurophysiology. As we see later in this chapter there is at least one set of arguments that suggests that that hope may not be attainable in the sense often expected by students of the mind–brain problem. We may have limits to our hopes for reduction to the attainable reality of description.

By *description*, we refer to a mathematical (or neural or computational) representation of the function of a system that maps the course of a process (i.e., its behavior), but in a way that is separable from assumptions about the specific internal physical or physiological mechanisms that might implement that function.

With this minilexicon in hand, it is now possible to continue on to the role of theory in a more general way. That is the purpose of the next section.

1.2. THE ROLE OF THEORY

If there is any single item that epitomizes the misunderstanding of science by the lay public (and sometimes by professional scientists) it is the imputed meaning of the word *theory*. To the layperson, a theory is a pejorative. It suggests ideas that are unproved and tenuous, ideas that should not be considered seriously because they are only the wildest speculation from people who have not yet "proven" their arguments. Theories, from the most extreme versions of this point of view, are unanchored in any way to reality and can be as deviant from the facts as the theorist is able to imagine. Inventive ingenuity is the only thing that places bounds on theories, according to this negative view. In this context, theory is equated with unsubstantiated speculation.

Theories, however, play an entirely different role in serious science. Because scientists appreciate that there can never be a complete proof of anything beyond the null hypothesis (the assertion that there is no relation between two events or two variables) and that data are always incomplete, the essence of progress in science is the gathering together of facts under as broad an umbrella of consistency as possible. To the degree that facts are not integrated into a theory, they remain impotent and without the possibility of generalization or extension beyond the miniature universe that spawned them.

The important point is that the cumulative combination of many facts into a theoretical overview is the essence of science. It is not an unanchored imaginative extrapolation (i.e., speculation), but rather a compilation of what is known in detail into a statement of what should be expected in general. The word *expected* is also central in this context, because a good theory should be able to quantify expectations by predicting what is likely to be forthcoming in situations that are similar, but not identical, to those that gave rise to the original data. How far afield predictions may be cast, of course, is a matter that is not specifiable in principle and can only be resolved in practice.

A theory such as the model presented in this book is, therefore, essentially a synthetic tool. Its role is to bring together data and observations into a coherent statement of what are thought to be the general principles governing a given domain. Contrary to the opinions of an ill-informed public, theories are not unanchored to reality; they dote and depend on the data and are successful only to the extent that they extrapolate from one set of specific measurements to others.

A theory must also, of course, be analytic to a certain degree. It should, within reasonable limits, allow us to take data from some lower levels of analysis and link them to a higher level. However, the degree to which a theory can be analytic or reductive between levels is much more highly constrained then is usually appreciated. This topic is treated extensively in another book (Uttal, 1998).

Given these strengths and constraints, what then is the role of an inductive, summarizing theory or model. The answer to this perplexity is simple and compelling: Theory is the raison d'être of the organized, formalized expression of our innate curiosity that we call science! It is the supreme goal of all scientific research. It is the ultimately desired contribution and the objective toward which all empirical work should be aimed. Theories are general, inclusive, and comprehensive statements of our understanding of the nature of the world we inhabit. Individual findings, facts, and data are useful only to the extent that they add to that general understanding. Gingerich's comment in the front material of this book captures exactly the flavor of the point made here.

Experiments, in this context, are merely local probes or assays that help us toward that general understanding. We conduct experiments to provide the grist for the theoretical mill; we do not concoct theories simply to provide a framework for a set of data. Data from experiments are rarely unique; they are all replaceable by some other experimental result. The theory-based understanding may in fact become unique as it converges on reality. Even prediction is not, or should not be, the goal. Although useful to asserting control over our world, it too is secondary.

The essential idea is that global understanding is primary; data and control are important but secondary! "Barefooted" empiricism, devoid of synthetic summary, is a sterile and empty endeavor. Mere data collection, without reason or conclusions, is an utter waste. Taxonomies can help by collecting, organizing, and cataloging, but even that role is secondary. The ultimate role of science is to summarize and surpass the accumulated data by stimulating general statements of understanding—in other words, to formulate theories that transcend observations. A theory is, at its most fundamental level, a statement of the universals for the domain of ideas with which it deals. The domain need not be all-inclusive; it may itself be a universe of limited extent. Nevertheless, any theory, even of a restricted domain, should strive for the most comprehensive statement of the nature of that domain as its ideal goal. Of course, the larger the domain encompassed by the theory, the better it is. This is the reason that unified grand theories are usually the most respected. It is not because they are more precise or accurate—they often are not. Simplicity arises out of breadth, not decimal points.

Thus, the role of the theoretical must never be minimized or ignored if the science is to be of the first quality. The role of the individual experiment is important, of course, but it is a truism that no particular experiment is essential. Should any particular experiment not be carried out, the loss would not be great. Almost certainly, another example of the same general point would subsequently become available. If no summary is created of a body of individually nonessential experiments, however, the loss is profound, because the goal of general understanding would not be achieved.

Theories come in many guises. Traditionally, experimental psychology crafted its theories out of the raw material of data summarized by descriptive statistics. Null hypotheses were generated and then tested to see if statistical differences could be detected that would allow their rejection. If the null hypothesis could be rejected, the hypothesis was tentatively accepted. Such hypotheses-driven theories had a tendency to be highly circumscribed and limited to a very restricted domain of inquiry—sometimes to only an experiment or two.

In recent years, theories and models of many different and more general kinds have evolved. Structural models have been developed that associate the behavioral and presumed mental functions by analogy with the operations of other devices. Computer programs serve as models of percepts (as well as of more complex cognitive processes) and neuroanatomical mechanisms. "Flow-chart" models are regularly invoked to simulate or describe cognitive processes. Mathematical models that differ greatly from the statistical ones of the past have been developed that are particularly relevant to the modeling of perceptual phenomena. Analogies drawn between mental and physical systems, a strategy that has a long history, are updated as each new technology or mathematical system emerges. Most recently, some of our colleagues have utilized concepts of chaos (e.g., Townsend, 1990; Skarda and Freeman, 1987) as models of brain and mind systems; others use the language of quantum mechanics (Bennett et al., 1989); and still others find that the metaphor of waves describes a wide variety of psychophysical judgments (Link, 1992).

Whatever the theoretical approach, it is clear that there has been a gradual evolution over the years from micromodels to more expansive macromodels. Although there is great diversity in the exact mechanics that is invoked, most interesting steps forward have tended to be more inclusive than was typical of only a few years ago. Whatever the scope, whatever the domain, the important thing is that human progress in acquiring knowledge and wisdom and solving problems depends mainly on our theoretical achievements. The rest (i.e., the solutions to the practical problems that face us) follows automatically—once basic understanding is in place.

The excursion into the philosophy of theories is presented with a specific motive in mind. Our goal is to justify the unusual breadth of the unified model presented in this book. We believe that by attacking the problem of vision on a broad front, we approach true understanding more effectively than if we had chosen to work on only a narrowly defined portion of the problem of how we see.

1.3 THE NEED FOR A UNIFIED THEORY

This section also emphasizes the argument for the broad-brush theoretical approach we have taken as we have developed the present vision system. That is, as noted earlier, all too many contemporary theories are severely restricted in their range. Rather than being broad, general statements of a wide variety of information, they tend to be narrow and restricted. Indeed, it is all too often the case that theories are based on a single set of data extracted from a single experiment. This is not theory making, in our opinion; it is just mindless curve fitting! Any theory worth its name should be inclusive of data coming from a spectrum of related experiments.

This is not just an aesthetic judgment. It is astonishing how often in experimental psychology, in particular, we have been misled by the results of a striking experiment to overgeneralize. That is, a single experiment may suggest a generalization that is found not to hold in later experiments or when that individual experiment is discovered to be flawed in some substantial way.

Nevertheless, there is also an aesthetic issue of great interest in this context. Vision science has moved ahead to a point where the single-parameter experiment and the single-experiment theory are really obsolete. Although these were necessary precursors to a more highly developed and modern multidimensional approach to theory, it now seems timely that we move on to both new empirical technologies and new theoretical strategies. Above all else, the beauty of a strategy that leads to models—more like the biological reality they seek to emulate and the fundamental truth they seek to represent—should not be denied.

Of course, there are harbingers of this new approach. Our laboratory (Uttal et al., 1992) has championed a unified theory of perception. Much of this present book is based on the development of that general approach. The late Allen Newell's (1990) effort to develop a unified theory of cognition was also a major step in this direction. Although neither of these efforts was exhaustive and complete in unifying all aspects of human mentation (strikingly illustrated by the fact that our work and Newell's did not overlap in terms of their substantive content), each explicitly moves away from the traditional microtheoretical approach. As we see in Chapter 3, there are many other examples in the field of perception that also are exploring the idea that a unified theory is possible, at least within the constraints of a reasonably broad subarea of experimental psychology.

1.4 ORGANIZATION OF THIS BOOK

This introductory chapter has spelled out some of the general philosophical perspectives and predilections that guide the development of this book. The remainder of the book deals with much more specific materials.

Chapter 2 provides some of the foundation material underlying our point of view. It presents a discussion of the physiological and psychophysical data that support the idea of separate channels and demarcatable centers.

Chapter 3 reviews various theories of combination, fusion, or binding. This chapter is a hybrid in which we review of some combination methods developed in other laboratories and present some original ideas from our laboratory.

Chapter 4 presents the overall structure and logic of our model. It also describes some of the technical details involved in carrying out this computational modeling project.

Chapter 5 presents a powerful new method for combining two-dimensional images based on gravitational physics. This work also shows how several different attributes can be combined to produce good object segmentation, or what perceptual scientists would call good separation of figure from ground.

Chapter 6 carries the idea of combination forward into the third dimension. Here is presented a scenario for combining three vastly different techniques for recovering surface shape—from shading, from structured light, and from stereo—into a much more robust process than represented by any one alone. Combination in this case permits us to recover shape information that would be ambiguous at best or irretrievable at worst.

Chapter 7 presents a much improved version of our shape recognition model. The program we describe has evolved to a point that it can recognize two-dimensional partial or occluded figures, thus imitating human perceptual closure and filling.

Chapter 8 describes another important part of our model. The reconstruction of three-dimensional surface shape from sparse samples taken over that surface.

NOTE

1. Though stereo blindness has been reported to be prevalent among binocular observers, many of these cases seem to result from inadequate strategic

registration of the two images rather than a true absence of stereo vision. In a study of this problem (Newhouse and Uttal, 1982) we found that nearly all observers could perceive depth in random-dot stereograms, given sufficient training and enough time to register the dichoptic images properly. Only two exceptions were encountered: one had a glass eye, and one had surgery for strabismus as a child!

2

Neural and Psychophysical Foundations of a Vision System

Channels and Centers, Interactions and Combinations[1]

2.1 INTRODUCTION

The theory of visual processing presented in this book is based on two fundamental premises. The first premise (I) asserts that the different kinds, or attributes, of sensory information are conveyed along different pathways, or channels, to localized nodes where different mechanisms transform these incoming signals. The second premise (II) argues that this functionally, if not anatomically, isolated information is subsequently combined or bound into a unified perceptual experience at some higher level of processing. Premise I is directly supported by a large corpus of anatomical and neurophysiological findings and is indirectly supported (within the constraints of reductionist theory discussed in Uttal, 1998) by certain psychophysical data that are, at least, "not inconsistent" with the idea of separate channels. It is the purpose of this chapter to review the physiological and psychophysical data that supports Premise I.

Premise II, on the other hand, is far more speculative, with only the barest initial empirical glimmerings suggesting the nature of the binding process. One possibility is that the segregated information representing the various attributes of the stimulus scene all converge on some common location in the brain where they are literally mixed and integrated. On the other hand, it may be that binding is accomplished by simultaneous or synchronous occurrence, perhaps involving some sort of a distributed signal to link distant events temporally. Binding, according to this alternative view, may simply be equivalent to neural activity in widely dispersed regions of the brain occurring within a limited time interval.

At the present time, there is no way to choose among the current and preliminary speculations concerning which of these two equally plausible possibilities is the correct one. Our later discussion of binding, in Chapter 3, is therefore much more theoretical and hypothetical than the more empirical and evidential discussions of data presented in this chapter that support premise I.

The concept of separate channels and nodes for different kinds of stimulus information *within* a given modality is completely consistent with, and is a logical extension of, established theory of sensory information processing. It is universally accepted that the sensory nervous system possesses separate channels for the different major modalities (audition, vision, somatosensation.) This general idea is supported by the simplest gross anatomical study of the nervous system; i.e., there exist optic, auditory, somatosensory, vestibular, olfactory, and gustatory pathways from the specialized, modality-specific receptors to the central nervous system. It is also supported, by easy demonstrations, that no matter how one stimulates any of the receptors, the perceived experience is always defined by the nature of the receptor rather than the nature of the physical energy. Indeed, this fact was long ago instantiated in the form of what was perhaps the first fundamental law of sensory information processing—Mueller's law of specific energy of nerves (1840). Mueller suggested that signals being conveyed along any sensory nerve produce perceptual responses that are *specific to that nerve*, no matter what physical energy, adequate or inadequate, was applied.

It is now also appreciated that at least some of the regions of the brain are also modality specific in the same way as are the afferent nerves. A century of research has shown that there exist specific regions or nodes in the cerebral cortex that are the primary receptive regions for these afferent nerves. Direct stimulation of these areas with an electrical stimulus produces only the auditory, visual, and somatosensory experience that are as modality specific as those generated by signals normally transduced by the proper adequate stimulus and communicated along the related sensory pathways. Shortly after signals leave these "primary" receiving areas, however, the situation becomes more complicated, and complex conscious experiences can be produced by electrical stimuli.

Figure 2.1 is a classic diagram showing the various cortical areas now believed to be associated with motor and sensory functions. Maps showing this kind of cortical localization of function are based on the work of a number of important neuroanatomists. Most notable among them in modern times are Penfield and Roberts (1959), Woolsey (1952), and the more recent studies of Kaas and Huerta (1988), Van Essen and his colleagues (1992), and Zeki (1993).

We now appreciate, however, that the classical concept of specific receiving areas and totally segregated cortical representation areas for the individual sen-

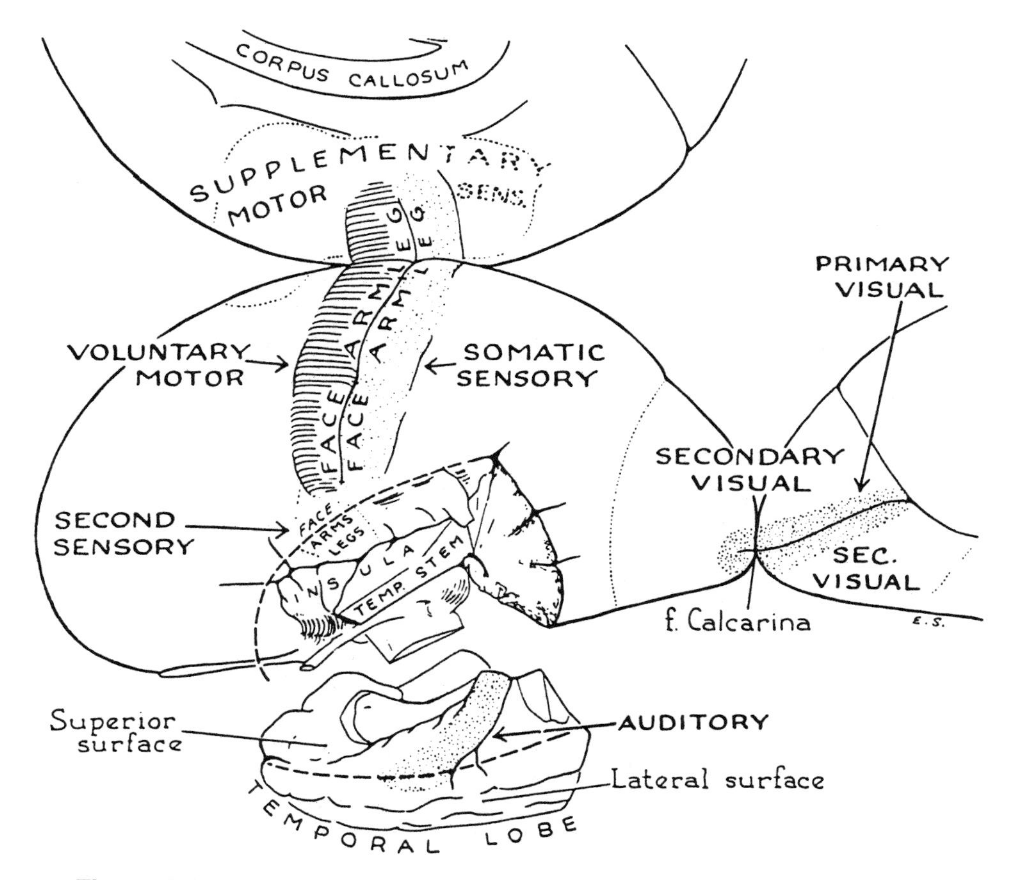

Figure 2.1 The classic view of the primary visual, somatosensory, and auditory sensory receiving areas and the motor areas of the human brain. In general, this model is still accepted although it is now appreciated that there are many additional visual and auditory regions not indicated in this figure. (From Uttal, 1978 after Penfield and Roberts, 1959.)

sory modalities is incomplete. Strong evidence supports the existence of multimodal regions of the central nervous system (see Stein and Meredith, 1993, for an complete survey of the multimodality of many cerebral areas). Some of this convergence apparently occurs very early along certain of the sensory channels. Stein and Meredith pay particular attention to the superior colliculus, long thought to be mainly a visual region but now appreciated to be a region containing neurons with sensitivities to several different stimulus modalities. The reticular formation of the brain stem, another early way station in the ascending nonspecific pathway, is also known to contain many neurons that are multimodal. This is not surprising, given that this system is believed to have evolved to mediate diffuse arousal or activation functions regardless of the modality that has been activated. For an extended discussion of the arousal role played by the reticular activating system, readers are directed to the classic work of Magoun (1954) and Moruzzi (1954).

At the same time, there are other regions in the visual pathway, such as the lateral geniculate nucleus of the thalamus, that do seem to be modality specific; that is, neurons in these specialized transmission centers convey sensory information from only a single receptor system, such as vision or hearing. The lateral geniculate nucleus (unlike the superior colliculus, which mediates spatiotemporal orientation and integration) contains neurons that are solely responsive to multiattribute visual inputs. At the same time, however, other, nearby regions of the thalamus convey information exclusively from other single senses. For example, only auditory signals are thought to pass through the inferior colliculus and the superior olivary nucleus.

This is the classic view of the molar organization of the ascending sensory pathways. (More detailed anatomies and discussions of this important set of topics can be found in Uttal, 1973, 1978.) Central to understanding the organization of the visual system, however, is the fact that although the lateral geniculate nucleus responds only to visual inputs, there is considerable segregation and separation of attribute information to be observed in this important visual relay station when we examine it at a more microscopic level.

With the exception of the three types of color receptors, the concept of microscopic segregation and separation of attribute information within the great visual modality, however, is definitely neither a classic nor yet even a universally accepted modern idea. Rather, it is a distinctly novel development, emerging only within the past decade. From some points of view, and hopefully without too much exaggeration, the idea of microchannel, or attribute, segregation may be as significant as Mueller's macroscopic law of specific nerve energies. It takes the notion of *specificity* down to another level of analysis, from the great modalities (i.e., vision, audition, somatosensation, etc.) to the individual attributes.

Visual science has now identified and conceptualized a system that seems to consist of distinct and separate channels for color, depth, texture, two- and three-dimensional form, and motion! In short, Mueller's law for the great sensory modalities can be extended down to corollaries that describe the level of specific sensory attributes. Each type of specific submodal information is carried along its own channel; in some cases, these channels can be identified by microelectrode-based electrophysiological experiments.

In principle, according to this hypothesis, if one could identify the physical locus of each channel and could stimulate each channel separately, the perceptual response would be specific to the subchannel, again regardless of the nature of the stimulus. In other words, if we could identify the specific channel for color, say, we could, in principle, generate a pure chromatic experience devoid, for example, of any motion or luminance experience. There are reasons why this cannot be achieved in practice, but the principle remains.

The very important idea of submodal, microscopic channel segregation embodied in premise I is an important and novel development in visual science. From some points of view it may be considered to be a paradigmatic shift of major magnitude in vision science. As usual, there were abundant clues in the older visual literature and experimental results that this form of organization might be the case. But as with so many other important ideas implicit in data, it was not until much later that the proper inferences were drawn. What we now appreciate to be the extension of Mueller's law to the attributes rather than the overall modality did not gain wide recognition until recently. We now review the scientific data supporting premise I that has emerged from research in the anatomical, physiological, and psychophysical laboratories. The general plan for the remainder of this chapter is to ascend the visual pathway from the retina to the highest cerebral stages. We consider the evidence for premise I as our discussion moves from one level to the next. Following the discussion of this anatomical and physiological data, we discuss how psychophysical findings provide some suggestive, if not definitive, support for the channel hypothesis.

2.2 CHANNELS IN THE PERIPHERY

2.2.1 The Main Visual Channels

Much of the information that we possess of the anatomical aspects of the visual system is based on observations on primates other than humans. For obvious reasons some experiments simply cannot be carried out on humans. It would be impossible, for example, to probe the living human brain for electrophysiological correlates of psychophysical responses in any systematic way. Even in some hypothetical situation in which the ethical constraints were not in place, physiological limits (e.g., the length of time a subject could remain conscious) would preclude a human psychobiological experiment. The use of animal models is, therefore, a necessary part of the process of accumulating information about the organization of any neural system.

The gross anatomy of the primate eye, and even the microscopic anatomy of the tissue contained within it, has been well understood for over a half century. At the optical microscopic levels, the works of Polyak (1941, 1957) and Rodieck (1973) are still important sources of knowledge about the structure of the neural elements contained in the retina. By using tissues and enucleated eyes from human cadavers, visual scientists have been able examine at least the gross morphology of the human visual nervous system, even if active neurophysiological results are rare. Substantial amounts of the information concerning the anatomy of our eyes comes from the examination of such tissue. Figure 2.2 is a drawing of the human eye, typical of most primate eyes. Most of the structures indicated on this diagram are functionally nonneural, in the sense that the visual information is still represented by the electromagnetic energy of light and the information processing is optical rather than neural. Rather than being involved

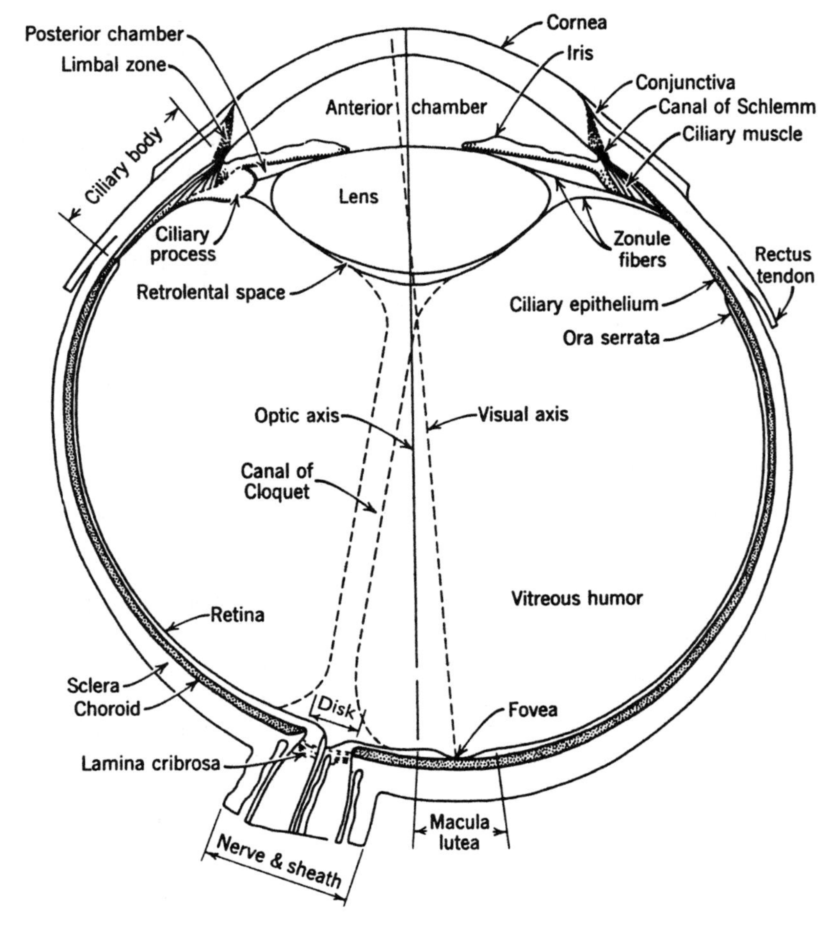

Figure 2.2 A diagrammatic sketch of the human eye. (From Uttal, 1973, after Brown, 1965, after Walls, 1942, as modified from Salzmann, 1912.)

in the neural transduction process (in which light is converted into the neural response) or the communication and integration one, most ocular tissues perform some supportive, metabolic, or physical optical role necessary for the sustenance of the eye or the focusing of the visual image. (See Uttal, 1981, for a complete discussion of the perceptual effects that can best be accounted for by the optics of eye and the physical environment prior to neural transduction.)

The precise point of interface between the world of physical light energy and the electrochemical energy of the nervous system is to be found in the retina, the microscopically thin tissue lining the interior of the eye. Figure 2.3 is a

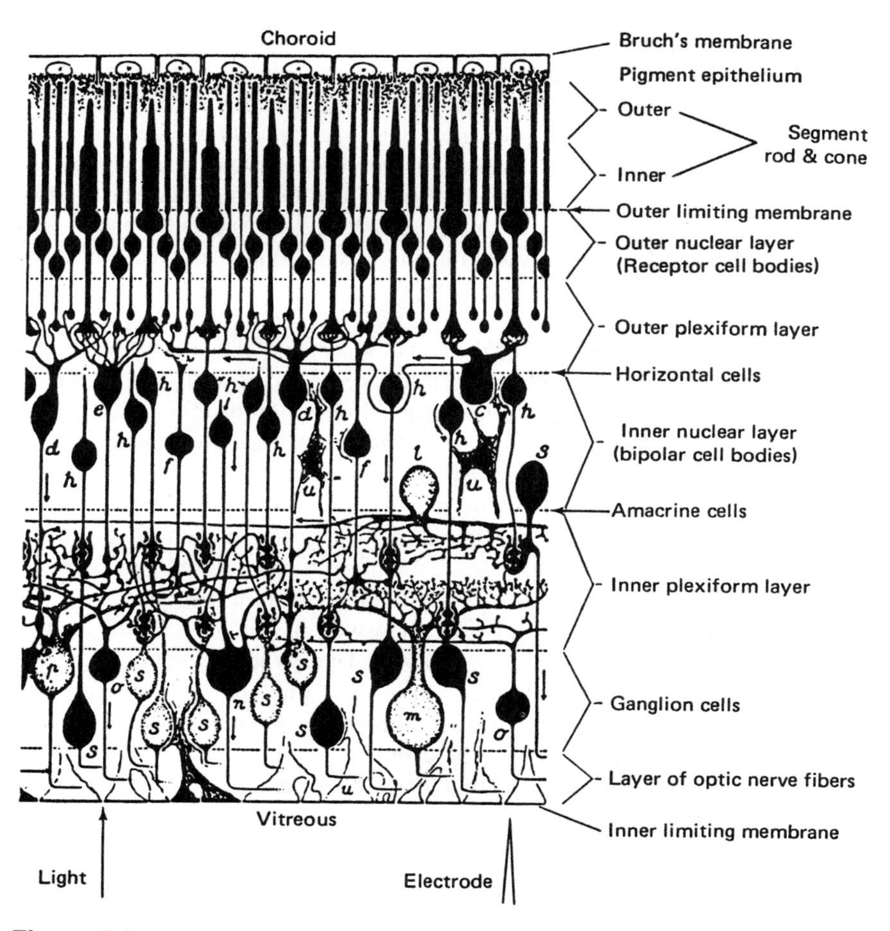

Figure 2.3 A diagrammatic sketch of the human retina. (From Uttal, 1973 after Ruch and Patton, 1965, as modified from Polyak, 1941.)

drawing showing the neuroanatomy of the retina. The retina is composed of three main layers of neurons: the receptors, the bipolars, and the ganglion neurons. At the two intersections between these three layers are two dense layers of synaptic interconnection: the outer plexiform layer and the inner plexiform layer. At these two layers of cellular interactions in the longitudinal chain of three neurons, there exist laterally connecting neurons of two other kinds. At the outer plexiform layer, neurons designated horizontal cells interconnect horizontally; at the inner plexiform layers, neurons designated Amacrine cells perform the same horizontal cross-connecting function.

The horizontal cross-connections between the receptors and the bipolar cells at the outer plexiform layer and between the bipolar cells and the ganglion cells, coupled with signal divergence and convergence within the ascending pathways, provide an enormous opportunity for neural information processing, recoding, and what is clearly various kinds of biological signal analyses. A wide variety of psychophysical and neurophysiological evidence supports the fact that these opportunities have been richly exploited by the visual nervous system. We consider this topic in greater detail later in this chapter. Our immediate task, however, is to consider the evidence for the separation of visual information into channels at the first level of transduction and coding—the visual receptors.

Receptor Channels

An appreciation of the anatomy of the visual receptors—the actual transducers of electromagnetic energy to electrochemical neural energy—is key to understanding the first stage of the multiple-channel operation of the visual system. The two types of photoreceptors—the rods and the cones—are shown in detail at a moderate level of electron microscopic magnification in Fig. 2.4. The anatomical differences between the two types of neuron that are apparent in this figure do not, however, explain this initial multichannel arrangement. To understand the signal separability at this level, it is necessary to consider the chemistry of the pigments that are contained within these receptor neurons.

Before examining the chemical story, however, it would be useful to state explicitly the nature of the channels we consider here. When the optics of the eye form an image on the retina, all of the attributes of any point on the image fall on a focused region of receptor neurons. From one point of view, the entire optical image may be considered to be essentially a single massive "channel," the analysis of which is carried out by the neural networks of subsequent stages of the visual nervous system.

The spatial arrangement of the image, the retinotopic map, is maintained up through several of the initial information-processing regions of the cerebral cortex, in spite of the analysis into the attribute channels. Unfortunately, some of the detail of the original image information may not be available because of the optics of the viewing situation. For example, image resolution may be impaired

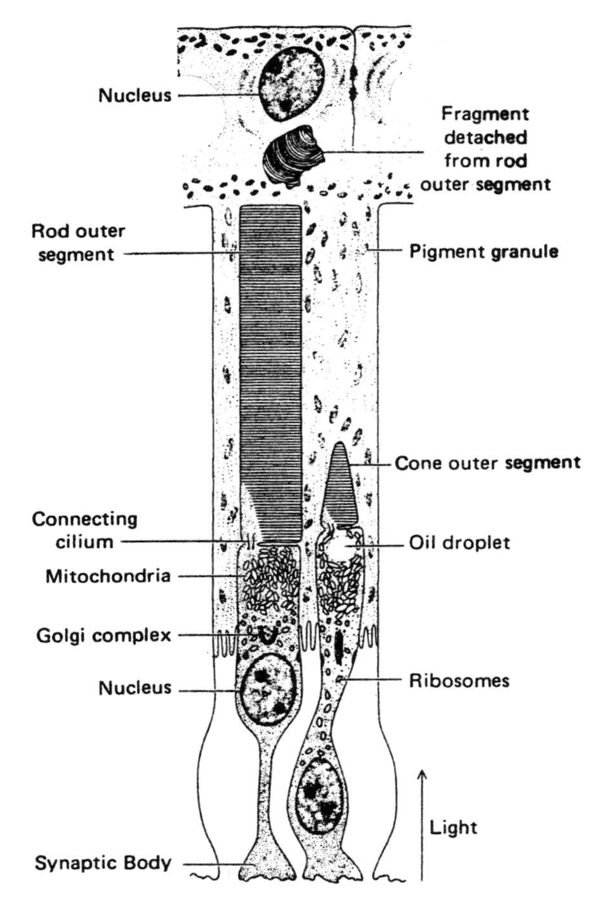

Figure 2.4 Diagrammatic sketches of the two types of receptors in the eye—the rods and the cones—in this case from a frog. (From Uttal, 1973, after Young, 1970.)

because of the optical aberrations of the eye. Some incident image information may be lost because of the limited range of sensitivity of the eye to either low luminances or limited spectral sensitivity; none of the four kinds of retinal receptors are sensitive to light outside the range of the visual spectrum—from about 400 to 760 nm. This does not mean that ultraviolet light and infrared light do not exist, only that they do not have any visual effect. Nevertheless, and even though it may be incomplete, a multidimensional (i.e., multiattribute) neural representation of the visual scene, to the extent detectable, is projected along the series of neurons that make up the ascending visual pathway toward the interpretive regions of the central nervous system.

The gross concept of an integrated single channel must be abandoned very early, literally at the point that the scene is transduced by the retinal receptors. In its place, we have to emphasize the fact that there is a separate microscopic channel for each point on the retina that can be neurally resolved. This change in point of view is based on the fact that the receptors are discrete, though microscopic, entities. Information about retinal locations is, therefore, maintained from the retina back to at least the cerebral cortex in a discrete rather than a continuous manner. Though distorted, a topologically consistent map (i.e., one that is an elastic transformation of the original retinal image) can nevertheless be measured on the surface of the primary receiving area in the occipital cerebral cortex.[2]

However, there is another, even more microscopic, system of channels than the retinotopic ones defining the specific loci of images on the retina. These channels conduct information about another attribute of the stimulus, its wavelength properties. Each location on the retina may, like the points on a television set, actually have several more microscopic attributes separately encoded within it. To expand on this concept of microscopic channels, two other levels of segregated attribute encoding must be considered. The first segregation is due to the "duplex" nature of the retina—there are two, quite different types of photoreceptor scattered across the retina.

The observed psychophysical differences (e.g., the shift in their respective peak sensitivities) between dim- and bright-light vision suggested the presence of at least two different visual channels: a scotopic one mediating a low-threshold black-and-white vision for the evening and night, and a photopic one mediating chromatic vision at higher-luminance, daylight levels. The anatomical basis of these two channels is the existence of the two types of anatomically distinguishable receptor neurons: the rods (mediating dim-light vision) and the cones (mediating bright-light vision). Rod vision is characterized by its achromatic, high-sensitivity, and low-spatial-resolution properties. Cone vision is characterized by its chromatic, high-threshold, and high-spatial-resolution properties.

The low-resolution properties of rods are largely accounted for by the relatively heavy convergence of rod signals onto higher-order neurons, not by any size difference between the two types of receptor. Thus, detailed information about spatial location is lost as many rods feed their signals into a common pathway. Conversely, this very same convergence is one of the factors that accounts for the high sensitivity of scotopic system to low light levels. In addition to the low threshold of the rod photochemical, the output of many rods can be pooled into a common, collectively more sensitive, system. The collective hypersensitivity is based on a statistical, noise-reduction process comparable to computerized averaging.

The fact that rods are the components of an achromatic system is accounted for by the fact that rods contain only a single photochemical with a single spectral absorption curve. Because of a relatively high concentration of this photochemical, the chemical nature of the substance, and the high degree of rod convergence, only a relatively few photons are necessary to excite a rod and, ultimately, the whole visual system. Indeed, the sensitivity of a rod, under optimum conditions, is so great that it seems that a single one can be excited in a visually effective way by a single photon (Hecht et al., 1942). It may even be possible for a full-fledged visual experience to be generated by a single quantum of light (Sakitt, 1972).

It is now accepted dogma that the *relative amount of activity* in each cone receptor is the specific information signaled to the brain that encodes the wavelength of the incident light.

Two kinds of psychophysical data have driven the two main alternative theories of color vision. Results from psychophysical color-mixing experiments were invoked by Helmholtz (1867) and used as the basis for what has been called the trichromatic theory. Color-mixing experiments showed that only three chromatic stimuli were necessary to reproduce any of the virtually infinite number of discriminable colors. The physiological-anatomical extrapolation of this idea was that there were three different kinds of photoreceptors. That is, in our present terminology, there are three distinguishable color channels differentially sensitive to portions of the wavelength spectrum of visible light.

Young, Helmholtz, and subsequent trichromatic theorists were very specific about the nature of these three receptor channels. These neurons, it was asserted, possessed spectral sensitivities such that they did not respond at all in total darkness and responded with progressively higher levels at suprathreshold-stimulus luminances, but with different spectral sensitivities. This model is trivariant and monophasic, in other words.

Hering's model (1878) proposed a much different type of sensitivity function for the receptors. His work, and that of subsequent scholars following his lead, was influenced mainly by a very different kind of color-mixing experiment, one in which spectral wavelengths were balanced against each other to cancel each other's chromatic properties—but without reducing the luminance to zero values. Thus, for example, a reddish-appearing light could be titrated, or added, to a greenish-appearing light until no reddish color experience was produced, just a yellowish or a bluish one. The reverse also seemed to work. Red could be canceled out by adding green. Yellow- and blue-appearing light seemed also to balance or oppose each other in the same way, leaving a reddish or greenish perceptual residue.

On the basis of this kind of experiment, Hering proposed that the receptors themselves were mainly opponent or biphasic in their action and consisted of three types of cones: a biphasic red-green opponent receptor, a biphasic yellow-blue opponent receptor, and a black-white receptor that simply followed the luminance of the stimulus but was not opponent. The black-white neuron was monophasic in the same sense as were the cone receptors proposed by Young and Helmholtz.

The central feature of the Hering opponent receptors was that, unlike the hypothetical Young and Helmholtz receptors, they were not quiescent in the dark, according to the Hering theory. Rather, in the absence of any stimulus, they would spontaneously respond at some intermediate level of activity. Their response could increase or decrease depending on the wavelength of the incident light. A typical set of Hering-type sensitivities is shown in Fig. 2.5.

An important aspect of the original controversy between the trichromatic and opponent theories of color is that it was only a debate about the physiological and photochemical nature of the receptors. The rest of the system was not a part of the controversy. There was no way to resolve what else might be happening at higher levels, since nothing was known of the neuroanatomy of the retina or the physiological responses of the neurons at high levels of the visual pathway.

It is also interesting to note that although they differed about the specifics, both of these theories proposed the existence of three receptor types. Thus, both would have been happy with the modern concept that the receptors represented and defined three different channels of information communication. Both theories are, therefore, variants of each other. Both accept the trivariant functional nature of the receptors, even though they differ with regard to the nature of the receptor's physiological and biochemical mechanisms.

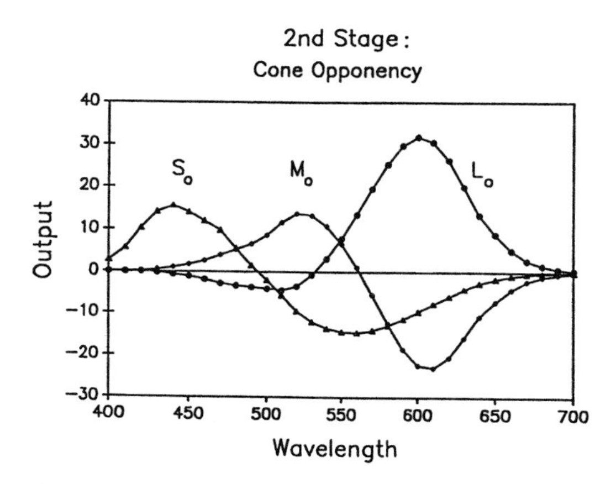

Figure 2.5 A typical set of Hering-type opponent spectral responses from the lateral geniculate body of the visual system. L_o, M_o, and S_o—respectively the long, medium, and short wavelength functions—would each have a comparable, but polarity-reversed, mirror-image equivalent function. (From DeValois and DeValois, 1993. Used with the permission of Pergamon Press.)

It is also appropriate to note that both theories are based solely on psychophysical evidence—the color-mixing experiments for Young and Helmholtz, and color complementarity (among other observations) for Hering. Therein lies the difficulty in resolving the original controversy. Psychophysical evidence is totally indeterminate with regard to the exact nature and sensitivities of these putative color channels. For years, although there was little debate about the trivariant nature of the chromatic channels, there was extensive debate concerning the receptor's exact sensitivities and the nature of the physical differences that accounted for both the trivariance and the results of the color-mixing experiments. Even within each of these disparate schools of thought, disagreements with regard to the details raged. For example, the width of the trichromatic spectral sensitivity curves of the three color receptors hypothesized by the Young–Helmholtz group was vigorously debated. Some (e.g., Hecht, 1931) argued that they were narrow and closely spaced; others (e.g., Judd, 1951) asserted with equal conviction that they were broad and widely separated.

Unfortunately, arguments over the exact shapes of the spectral absorption curves of the receptors could not be resolved at the time. There is nothing in the psychophysical data that can possibly distinguish between the alternate theories. Not only were just three colors necessary to reproduce any other color, but, even more surprisingly, virtually *any* three colors could be used. It was only when direct spectrophotometric measurements were made in the 1960s, virtually simultaneously by two groups (Brown and Wald, 1964; Marks et al., 1964), that the matter of the specific spectral sensitivities of the visual receptors was resolved and the nature of the early chromatic channels defined.

Figure 2.6 illustrates the current best answer to the continuing debate about the exact spectral sensitivities of the three cones. Three widely spaced but overlapping spectral sensitivities initiate the three separate channels of chromatic information, with peaks at approximately 445, 535, and 570 nm. These curves were obtained by measuring the absorption spectra of a sampling of human cones from enucleated cadaver eyes.

We now appreciate the magnificent synergy between the quite separate domains of analysis in developing our understanding of the receptor channels. First, psychophysics tells us that only three colors are necessary to match any other. Second, spectrophotometry tells us that there are three type of absorption spectra to be measured in cones. The latter information also allows us to resolve the longstanding debate between the Young–Helmholtz and the Hering schools concerning the exact physiological nature of the receptors. With regard to the receptors, Young and Helmholtz were correct. That much is now certain. Nevertheless, Hering was not entirely wrong. He was, instead, describing a neural language—a code—that was eventually to be found in the next layers in the retina, beyond the receptors. The important point is that both neural codes seemed to have left their trace in the final output of the system—the psycho-

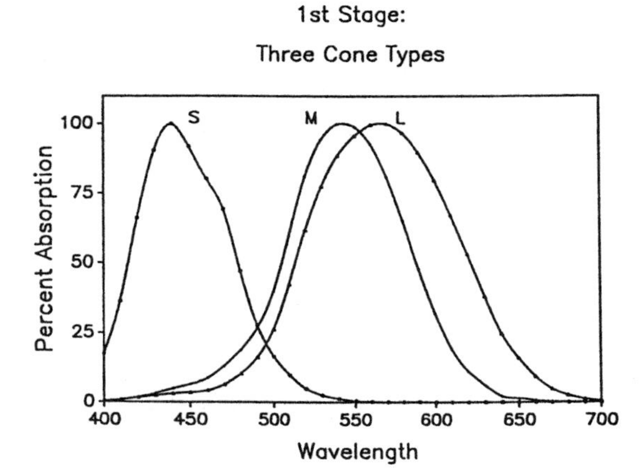

Figure 2.6 A current estimate of the spectral sensitivities of the three cones. (From DeValois and DeValois, 1993). Used with the permission of Pergamon Press.)

physical response. Although we cannot reconstruct the exact codes used at each anatomical level from psychophysical data alone, the perceptual trace can suggest what may have happened some place in the visual pathway. Which particular coded representation is tapped by a psychophysical experiment depends on the type of experimental procedure utilized. However, it is important to keep one fundamental point clear: Psychophysics, the overall measure of system performance, can never resolve a question of the internal mechanisms of neural coding or structural organization.

What was resolved at this point by the microspectrophotometrists was sufficient to satisfy neurophysiologists about the nature of the receptors. However, further investigations were required in another, even more microscopic, domain to provide a biochemical answer to the next obvious question: What biological or chemical property of the cones and rods accounts for their four different spectral absorption curves?

The answer to this question was found first for rods, in the photochemical laboratory of the distinguished Nobelist George Wald. Wald had been following up some 19th century studies by Boll, who had discovered a curious chemical in the rods of the retina. Boll called this substance *rhodopsin* and noted that it seemed to be involved in the visual process in some way. When extracted from rods in the dark, rhodopsin had a marked purple color; but when exposed to light, the material bleached to a yellowish color. Based on these observations and on some of the same psychophysical data we have already discussed, Hecht (1931) suggested a theory of dim-light visual transduction in which the rhodopsin was involved in a cycle of breakdown and regeneration.

It was in Wald's laboratory, however, that the critical chemical discovery was made: There was another substance present in significant amount in the rods. This substance was *retinal*, the oxidized breakdown product of vitamin A, the latter substance being otherwise known as retinol. On the basis of this discovery, Wald carried out the Nobel award-winning work in which he specified the exact sequence of the breakdown and regeneration cycle of rhodopsin under the influence of light and dark. Under the influence of light, he showed that rhodopsin breaks down into retinal and a large protein molecule he called *opsin*.

As rhodopsin breaks down, it passes very rapidly (in micro-and milliseconds) through a number of intermediate stages. These intermediate breakdown products could be individually studied by varying the temperature of the solution in which they were suspended, a technique known as *freeze chemistry*.

Rhodopsin, the rod photosensitive substance, is a large protein molecule made up of two parts: a large protein component (or moiety) and the smaller organic molecular component, retinal. It is important to appreciate that it is the retinal that is the light-sensitive component of this combined molecule, not the opsin component. Nevertheless, the complex chemistry of the process has many interactions and secondary effects that have made the search for an understanding of the visual transduction process one of the major challenges of modern biochemistry.

What we now know is that opsin and retinal are solidly bound together into a stable molecule only when the retinal is configured into one particular of its several possible stereoisomeric forms—specifically the 11-cis form of the molecule. When we say that retinal is light sensitive, we really are alluding to the fact a quantum of light may be absorbed by this molecular component. Even more particularly, it is absorbed at the eleventh carbon-to-carbon bond, the point at which the 11-cis molecule is bent.

Because of its relationship with the opsin molecule, retinal absorbs light with a specific sensitivity that varies across the visible spectrum. Surprisingly, the specific absorption spectrum of retinal is defined, not by the absorption characteristics of retinal, but by slight differences in the respective opsins in the different photosensitive molecules. This fact means that even though a single form of retinal is common to all four photoreceptors (the rods and the three cones) and even though the absorption of the quantum of light occurs in the retinal molecule, the absorption spectrum can vary among the four photosensitive substances.

The effect of the absorption of light by 11-cis retinal is eventually (after it passes through the intermediary stages identified with the freeze chemistry process) to transform it to all-trans retinal. All-trans retinal differs from 11-cis retinal in one very important function characteristic: It does not form a stable combination with opsin. Thus, when a quantum of light is absorbed and the stereoisomeric form of the rhodopsin molecule changes, it spontaneously starts to break up. How far the spontaneous breakup goes through the series of breakdown products depends on the temperature of the cellular environment. Thus by applying the controlled temperatures of freeze chemistry, the spontaneous breakdown can be studied at leisure, even though the normal-body-temperature–driven process would appear to occur almost instantaneously.

The reconfiguration of the receptor chemical produces a major change in the chemistry inside the cytoplasm of the rod receptor. In the dark, there are very large amounts of the substance known as *cyclic GMP* (cGMP). Its role in the visual process is to regulate the amount of Na^+ (sodium ions) that pass through the outer cell membrane of the rod. The larger the amounts of cGMP, the larger the flow of Na^+ through the outer cell membrane of the receptor neuron. The effect of the breakdown of the rhodopsin molecule (specifically the presence of increased amounts of free opsin) is to reduce greatly the amount of cGMP in the internal environment of the receptor. This occurs because of an intermediate step in which the concentration of the enzyme photodiesterase is increased by the free opsin. The net effect of this series of chemical events is that when light quanta are absorbed, the sodium permeability of the outer cell membrane of the rod is reduced. This has the result of increasing the voltage across the membrane. It hyperpolarizes (increases) the transmembrane potentials to values that are larger than normal.

The increase in membrane potential that thus results from the absorption of a light quantum by the retinal portion of rhodopsin is the final step in the transduction process. The electromagnetic energy of the quantum has been transformed into a coded representation that is instantiated in a totally different energy—the electrochemical energy of neurons. The transduction process performed by the receptor is now complete. The increase in the membrane potential is now available as a signal that electrotonically diffuses to the synaptic regions of the rod located at the outer plexiform layer. The transient increase in membrane potential, therefore, is the critical neural event that stimulates those synapses to release their transmitter substances, which, in turn, activate the horizontal, bipolar, and, according to some authors, other receptor neurons as well.

Our particular interest here, however, is not with the fascinating story of the chemistry of the transduction process per se, but rather with the reasons that there are physiologically differentiable channels in the receptor layer of the retina. The key reason that there are four channels (one rod and three kinds of cone) in the receptor layer of the retina is that there are four different opsins, one in each of the four receptors. Each of these opsins is slightly different than the other three in its chemical makeup (although it is not yet exactly certain where in the opsin is the critical portion that changes the retinal's spectral absorption sensitivity.) Each of these four different opsins combines with the single kind of retinal to produce four slightly different photochemically sensitive substances with different spectral sensitivities. It is this difference in the four retinal's spectral sensitivities (mediated by the differences in the chemical structure of the four opsins) that creates what we now appreciate are four different channels in the retinal receptor layer.

The story just told may not be totally complete, however. Very recent evidence (Neitz, Neitz, and Jacobs, 1991) suggests that although there may indeed be three different kinds of cones, there may actually be five different kinds of cone chemicals. The long-wavelength-sensitive cone spectral absorption curves and the medium-wavelength-sensitive cones may each actually be the composite results of two distinguishable cone chemicals with very closely spaced absorption spectra.

The end result, nevertheless, is that the information about color has been segregated into different channels with different spectral sensitivities. These different channels are interspersed among each other; but if one probes with a microelectrode or a microspectrograph, individual receptors can be identified with these distinct absorption spectra. The key conceptual fact and the basis of the combinability hypothesis—premise II, introduced at the beginning of this chapter—is that this information, segregated as it is in the receptor layer, must be integrated later. Specifically, the relative amount of activity in each neuron can be compared (combined, integrated, bound) by the central nervous system ultimately to produce discriminable color experiences.

But there are still some surprises ahead. First, it is important to remember that the very same neuronal channels that carry color information also carry other kinds of information about the two- and three-dimensional shape, movement, contrast, position, and texture attributes of the stimulus. Indeed, the three cones and the single rod also play critical roles in the function of other channels. The channels have not been completely separated at the receptor level. Indeed, they may not be totally anatomically distinct anyplace in the ascending pathway.

Our understanding of what is meant by a channel at the receptor level is, however, furthered by another important aspect of the three color channels. The trichromatically encoded information that their activity represents is almost immediately transformed into quite a different neural code that much more closely resembles Hering's original ideas regarding opponent receptors. We now discuss the modern view of this transformation from a trichromatic color system to several cascaded opponent color systems.

Chromatic Recoding at Higher Retinal and Central Levels

Two streams of information suggest that the color-coding scheme used by the whole visual system is not as simple as the trichromatic model of the receptor postulated by the Young–Helmholtz theory. First, psychophysical evidence has for a century (consider the date of Hering's work) exhibited some curious anomalies that do not fit with the simple notion of trichromacy. For example, it had long been known that there were combinations of colors—complementary colors—that canceled each other out when mixed in appropriate amounts. Jameson and Hurvich (1955), in what certainly must be considered to be one of the classic papers on this topic, also pointed out that some colors simply were not used together in any language. For example there was no word to describe a reddish-green light or a bluish-yellow light. This was so in spite of the fact that a wealth of color terms were available to describe the yellow-green and red-blue color experiences.

Important psychophysical experiments carried out by Hurvich and Jameson (1957) measured the amount of the opponent color (red in the case of green; yellow in the case of blue; and vice versa for both cases) that had to be mixed with a color to cancel out its chromatic experience. These data depict a chromatic sensitivity that goes well beyond the linguistic observations and that would be difficult to describe in conventional trichromatic terms. Although these results cannot tell us specifically and uniquely what the underlying mechanism is, they do suggest that something other than trichromatic coding is happening somewhere in the nervous system. Just how early in the visual system this recoding occurs did turn out to be a surprise—not the fact that it was happening someplace. It was actually for the second type of neuron in the retina—the bipolars—that substantial evidence exists of a transformation of the code for color.

The other main stream of work suggesting that there was more to the neural codes for chromatic experience than simple trichromacy was based on electrophysiological experiments, mainly those pioneering studies by Russell DeValois and his colleagues (DeValois et al., 1964, 1966; DeValois and Jacobs, 1984) in the lateral geniculate body.

DeValois and his colleagues reported complicated sets of response patterns in the lateral geniculate body. Although some of the geniculate neurons produced monophasic signals that seemed to vary with the luminance of the stimulus, many other of the neurons at this level also exhibited that same kind of wavelength-dependent de- and hyperpolarization that are now called opponent responses.

At this point we see that there are both psychophysical and neurophysiological pieces of evidence to support the notion that somewhere in the nervous system there is something other than a simple trichromatic code occurring. The discrepancy at this point is that the two sets of data do not agree in detail. To explain how this discrepancy might be resolved, we now turn to a discussion of the recent work of DeValois and DeValois (1993).

DeValois and DeValois proposed a modern version of the original zone, state, or level theory first suggested by such workers as Schroedinger (1925) and Adams (1923) but brought to its first modern form by Hurvich and Jameson (1957). A zone or stage theory is a physiological model that assumes that both trichromatic and opponent coding processes are present in the visual pathway but that somehow one (the trichromatic code) is converted into the other (the opponent code) at some higher synaptic level. Since the basis for all of these physiological hypotheses was originally purely psychophysical, there was no way in which the details of the conversion mechanism could be explained until much later in the history of our science.

The DeValois version of a zone or stage theory assumes that the trichromatic code used at the initial level of the three photoreceptors is transformed into an opponent code as early as the retinal bipolar cells by interactions in the outer plexiform layer. Therefore, both the nature of the codes and, of immediate interest in the context of the present discussion, the channels change from one neuronal level to the next. Whereas three different photoreceptor chemicals defined three different channels of information in the receptor layer of the retina, the stage or zone theory proposed by the DeValois asserts that the receptor outputs are combined by bipolar cells into at least eight distinguishable channels of information. Six of these, communicated along the midget bipolar to the midget ganglion neuron to the parvocellular layers of the lateral geniculate nucleus, are opponent-type color neurons. The different response sensitivities of these neurons are produced by various combinations of the outputs of the long- (L), medium- (M), and short- (S) -wavelength-sensitive cones. Specifically, they proposed that there are three types of opponent neurons that are excited by the long-, medium-, and short-wavelength regions and inhibited in other portions of the visual spectrum. The sensitivities of these three neurons are formed by subtracting the response of the sum of the three cone outputs (represented by $L+M+S$) from the output of the respective individual cones. DeValois and DeValois represent this kind of response sensitivity of three kinds of midget bipolar (i.e., three channels) by the following three formula:

$$L_o = L - (L + M + S)$$
$$M_o = M - (L + M + S)$$
$$S_o = S - (L + M + S)$$

The minus sign in this case means that the summed activity of "$L+M+S$" inside the parentheses has an inhibitory impact on the midget bipolar cell. The direct, unsummed, positive cone input (e.g., "M") has an excitatory influence. The left-hand side of each equation (e.g., L_o) is nothing more than a shorthand tag for the particular type of neuron whose response is being defined by that equation. The subscript "o" in this case refers to an opponent type neuron, just as the subscript "n", to be used later, refers to a nonopponent neuron.

In addition to these three kinds of channel, the DeValois model goes on to suggest there are three other channels embodied by midget bipolar neurons at this level. The responses of these channels are defined by inhibitory influences from the individual cones and excitation from the combined sum of their outputs, the reverse of the other three types. These three neurons would have mirror-image sensitivities to those shown in Fig. 2.5 and would be represented by the following equations:

$$-L_o = -L + (L + M + S)$$
$$-M_o = -M + (L + M + S)$$
$$-S_o = -S + (L + M + S)$$

In addition, there are also at least two types of nonopponent bipolar cells that convey overall luminance information. These two channels consist of bipolar cells that receive information from both the long-wavelength-sensitive (L) and the medium-wavelength-sensitive (M) as well as from the summed output of all three. Using DeValois and DeValois' convention of representing these interactions with equations, they may be represented by the following two expressions:

$$LM_n = L + M - (L + M + S)$$
$$-LM_n = -L - M + (L + M + S)$$

The important point for our discussion is that the three channels of the receptor layer have been re-encoded as at least eight channels in later parts of the chain of neurons. Each of these channels is in principle anatomically and physiologically distinguishable, according to DeValois and DeValois. Anatomically, they believe that all of these neurons in the second level of retinal anatomy and, for that matter, the second level of neural encoding are diffuse bipolar neurons sending information to the parasol ganglion cells and from thence to the magnocellular layers of the lateral geniculate body. This portion of their theory is directly implied by the physiological work described earlier.

DeValois and DeValois then went on to resolve another difficulty that arises when the calculated sensitivities of their model are compared to the pattern of the Hurvich and Jameson psychophysical responses. They note that the fit is not particularly good. On the basis of this psychophysical discrepancy, they propose that there is another level of encoding and that other channel arrangements are subsequently established in the visual pathway that convey the same information as did the receptor trichromatic system and the first opponent system, but in a different neural language or code.

According to DeValois and DeValois, it is the behavior of this third level of channels communicating color information that is compatible with the psychophysical data obtained from the color-cancellation experiments of Hurvich and Jameson (1955). Indeed, it was the need to establish this compatibility that led them to propose the existence of this third level of channels. Their model suggests that this transformation probably occurs among the parvocellular areas of the lateral geniculate nucleus, although this cannot be certain, since these neurons have not been physiologically identified. DeValois and DeValois suggest that Hurvich and Jameson's psychophysically defined opponent color mechanisms (red-green and blue-yellow) are created by the combination of the outputs of the M_o and L_o systems and modulated by the S_o system. Thus, according to their model, both the third-level *RG* and *BY* opponent-sensitivity curves matching the psychophysical data are produced in the following manner:

$$RG = L_o - M_o + S_o$$

$$BY = M_o - L_o + S_o$$

There are, in addition, two other possible channels ($L_o + M_o$ and $M_o + L_o$) that are essentially achromatic and, in principle, should be anatomically distinguishable from the *RG* and *BY* systems. Thus, with these four third-level and four magnocellular, achromatic channels, there are at least eight hypothetical channels of color information leaving the lateral geniculate body.

It is important to appreciate that the DeValois model is based on their felt need to show a kind of isomorphism between the performance of neural "channels" and the psychophysical data. At a fundamental philosophical level, however, isomorphism is probably not necessary, since all three levels represent the same information. There is no "best code." Unfortunately, isomorphism is an often-misused concept in the coding studies, as discussed elsewhere (Uttal, 1973, 1998). Furthermore, it is also important to reiterate an essential fact: Although the second level is based on actual neurophysiological findings, the putative third level is purely hypothetical. The third level is a hypothesis that from one point of view can be considered nothing more (or less) than a description of the psychophysical data in neural terminology. The possibility that some additional, even more subtle, encoding has taken place is high and cannot be overlooked. Any descriptive model of this kind must be neutral with regard to the actual physiology of the system.

This expanded discussion of one theory of color coding and channels in the afferent visual pathway should help to make clear what is meant by premise I. Many of the proposed channels that convey chromatic information are anatomically distinct entities. That is the fundamental fact attested to by a substantial body of electrophysiological evidence from the first two levels and suggested in the formulation of the third by the DeValois and DeValois theory. Though the nature of the channels may change and the language or code used by the nervous system may be transformed as signals pass from one synaptic junction to the next, there are many physiologically identifiable and anatomically distinguishable structures conveying different attribute information. Thus, the concept of channels seems to be a solid foundation on which to build either a theory or an actual artificial vision system.

However certain we may be about this essential foundation of premise I, it is also important to make another very important and closely related point: Some of these very same neurons may be carrying information about attributes other than color. For example, the midget ganglion cells that convey chromatic information also carry spatial information that will ultimately be interpreted as the two-dimensional shape of a stimulus object. The triad of color receptors in local regions are embedded in the interstices of a global retinotopic map that represents, with high resolution, the two-dimensionally projected shape of objects in the visual field. This type of multiple use of single neurons may be designated as *overlapping coding*. Overlapping coding thus refers to the idea that single anatomic channels may convey information about multiple attributes of a stimulus. A single neuron, therefore, may convey information about shape and color (or motion and contrast) because of its dual role in the local and global components of a highly interconnected system.

Thus, the three types of cones, segregated as they are for wavelength encoding, are also the overlapping channels for precision spatial vision. The rod-free region of the central retina—the fovea, where cones are concentrated in the highest density—is also the region of the highest spatial acuity. Therefore, even though the chemicals contained in the cones determine their roles as chromatic encoders, the way they project collectively from the outer plexiform layer to higher levels determines their roles as encoders of spatial patterns.

To make this point more specific: The cones connect to midget bipolars, which in turn connect to midget ganglion cells, which in turn connect to the parvocellular regions of the lateral geniculate body. This train of "midget neurons" is anatomically characterized by the limited spread of their dendritic connections. In other words, the information received by a single cone is conducted centrally, with a relatively small amount of network dispersion, thus maintaining the high spatial resolution of the system.

Rods, on the other hand, have a much higher degree of convergence; i.e., many rods connect to a single bipolar, which in turn converge on the widely spread dendritic umbrella of the parasol ganglion cells. By the time these neurons have arrived at the magnocellular regions of the lateral geniculate body, there has been massive overlap and the identity of particular locations on the retina has been lost. This anatomic arrangement accounts for the low spatial acuity of the magnocellular regions.

The relatively large amount of convergence from single rods does, however, produce a high sensitivity to very small amounts of light and good contrast sensitivity. This can be accounted for by the accumulation of stimulus information from a relatively large region of the retina. The statistics of this situation permit the identification of a small amount of light as a valid external stimulus and not a random or endogenously generated response.

The important point is that different attributes of the visual system sometimes may be in conflict with each other. Good spatial resolution and good intensity sensitivity are competitive processes. The process of evolution has resolved this competition of needs by developing two different channels—the parvo- and the magnocellular pathways—so that both visual needs can be satisfied in a single nervous system. The development of alternate channels for the different attributes of a visual scene may be evolutionarily forced by just this kind of conflict of coding requirements.

The converse of overlapping coding, *redundant coding*, is also evident in the color system we have described. Redundant coding refers to the fact that two or more physiologically and/or anatomically identifiable channels may carry the same information. Color information need not, according to this idea, be carried by any particular code at any level. Rather, as the work of DeValois and DeValois has shown, there are often many types of neurons redundantly coding the same information, with quite different patterns of response—i.e., different neural codes. DeValois and DeValois' (1993) theoretical model suggests six or seven or eight different types of opponent neurons, any one of which may be functionally redundant with any of the others as well as with the early trichromatic code.

The important thing to keep in mind in this context is that the system is still trivariant. The information initiated by the three cone receptors may be encoded and recoded into many more than three different types of neuronal channels. Nevertheless, the *information*, no matter how redundant the coding system may be, is still defined by the fundamental trivariance of the receptors *and* all other coding levels. In short, additional channels do not necessarily mean additional degrees of freedom, merely redundant ways to encode the same information.

The Magnocellular and Parvocellular Geniculate Channels

Let us now consider the arrangement of the peripheral channels in greater detail. As the anatomy of the retina became more clearly understood in recent years, it was observed that there were at least two major classes of retinal ganglion cell forming the third layer of the retina. In addition to certain physiological distinctions, these two classes of neuron differed in terms of the structure of their dendritic arborizations. One type, collectively displaying larger arborizations, has been tagged as the parasol or diffuse ganglion neurons The other smaller type is now referred to as midget ganglion neurons. Each of these types of ganglion cells receives information from a comparable bipolar, a diffuse bipolar or a midget bipolar, respectively.

Although it was not at first appreciated that there was any significant functional distinctions between the two anatomical types of ganglion cell, it is now understood that they also represent the elements of two different functional channels to the central visual nervous system. In fact, the information conveyed by the midget and parasol ganglion cells, respectively, is immediately segregated as these neurons project to their first synaptic connection at the next neuronal relay station, the lateral geniculate nucleus of the thalamus. The parasol ganglion cells project to regions of the lateral geniculate nucleus where the cell bodies are much larger than the regions to which the midget ganglion cells project. Figure 2.7 (from Livingstone and Hubel, 1988) shows the segregation that occurs at this point. The midget ganglion cells from each eye project to two of the four dorsal *parvocellular* (i.e., small neuron) layers, although the larger parasol ganglion cells project to one of the two ventral *magnocellular* (i.e., large neuron) layers. (The other three layers receive the equivalent signals from the other eye.)

The analysis of the organization of the parvo- and magnocellular regions by Livingstone and Hubel (1988) was one of the main stimuli to the development of contemporary theories of the segregated-channel notion that we have called premise I. The critical idea in their analysis was the idea that the magno- and parvocellular regions are anatomically distinguishable pathways whose identity is maintained up to the primary visual receiving area (V1, or Broadman's Area 17) of the cerebral cortex. They also suggested that not only were these anatomically (i.e., by neuron type) defined channels anatomically distinct, but they also encoded and conveyed different attributes of the visual stimulus. For example, the parvocellular channels that pass through the four dorsal layers of the lateral geniculate body convey information that is decoded to discriminate colors and

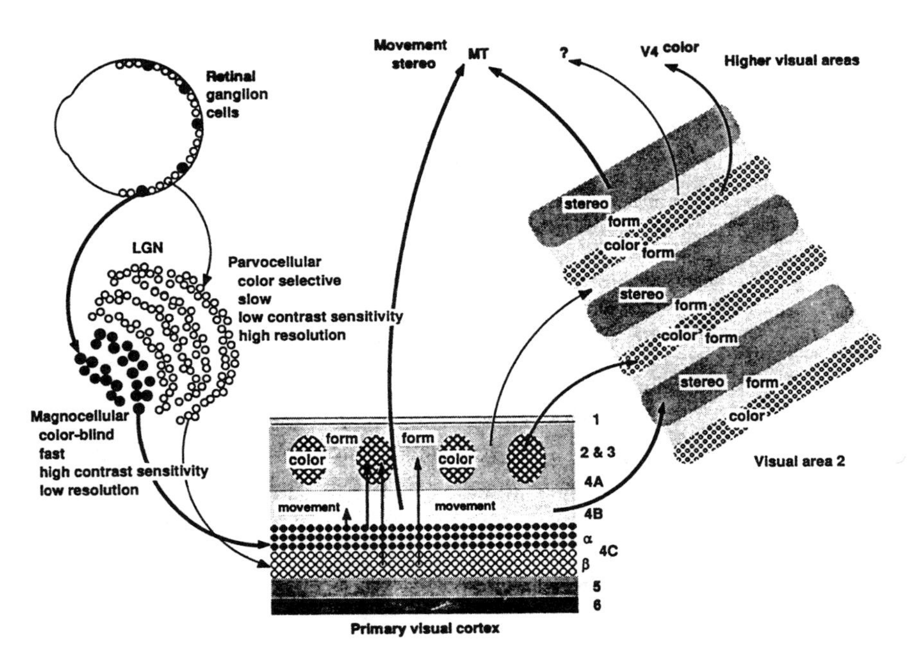

Figure 2.7 The segregation of the several channels for different attributes of a visual stimulus. (From Livingstone and Hubel, 1988. Used with the permission of the American Association for the Advancement of Science.)

to resolve closely spaced objects. However, the temporal properties of this channel are poor and do not encode contrast discrimination adequately. On the other hand, the magnocellular channels that pass through the two ventral layers of the lateral geniculate nucleus have good temporal responses and high contrast sensitivity, but can only poorly convey information necessary to resolve objects and distinguish color.

The nature of these two pathways provides another piece of evidence that helps us to understand the way in which stimulus attribute information is partitioned and sent along what are in this case well documented, anatomically distinguishable channels. We can only speculate why this type of system evolved, but there is no question that functional segregation into specialized channels occurs in the visual system. As we now report, this type of organization is also maintained up to and including the highest reaches of the cerebral cortex.

2.2.2 Other Visual Channels

So far, we have dealt with only one set of visual channels. It is known, however, that information from the retina is projected to several other portions of the cerebral cortex than just the primary receiving area (V1). The main visual pathway, without doubt, is the geniculo-striate pathway that we have been discussing: retina → lateral geniculate → occipital (striate, primary, V1) cerebral region. However, a substantial body of evidence suggests that there are other pathways and, therefore, other channels that carry visual information to the central nervous system.

The Pretectal Pathway

In emphasizing the geniculo-striate pathway so far, we have concentrated on the processes of vision that lead to our awareness of the color, shape, texture, depth, or movement of a scene. Information acquired by the eye, however, is used for other functions beyond scene perception. Alternative uses of the incident-light information is now believed to be mediated by channels that are quite separate from the geniculo-striate pathway. One of these alternate channels projects to the pretectal area of the brain stem. The pretectal area is the integrative center of a reflex system that controls the diameter of the pupil, probably without cerebral (conscious?) intervention. Pupil diameter is primarily, but not entirely, a reflex function driven by the luminance of the stimulus. It is also a function of some cognitive feedback signals, autonomic responses, and the accommodative state of the eye.

Efferent signals from the pretectal area pass directly to the accessory occulomotor nucleus and then to the ciliary ganglion that controls the ciliary muscles—the mechanical effectors varying the diameter of the pupil. The pretectal pathway or channel conveys information in a way that seems to be completely independent of the lateral geniculate nucleus. This alternative channel branches off the optic tracts prior to their entry into that thalamic center.

The Superior Collicular Pathway

Also branching off from the optic tract prior to the lateral geniculate nucleus is a pathway that goes to the superior colliculus of the midbrain. The superior colliculus acts as an integration center for signals from the eyes, the ears, and postural receptors. Its main function is to control reflexively the movement of the eyes, to help in ocular stabilization and orientation. These signals are combined and feed directly into another layer of the superior colliculus, which is a motor control region. This form of eye movement control also occurs without cerebral mediation and can also be considered to be a reflex circuit. (The movement of the eyes, of course, can also be intentionally controlled by conscious effort.) In many ways the motor-coordinating role of the superior colliculus for the eyes seems comparable to that of the cerebellum for the rest of the body.

Some time ago there was a rather surprising result that suggested that the function of the superior colliculus, or, as it is otherwise known, retinotectal, channel might be more complex than a "simple" reflexive motor-orientating system. Weiskrantz et al. (1974) demonstrated what they believed was residual visual capabilities in people who had been partially decorticated in one of their primary visual regions. If a stimulus was placed in the region of the visual field projecting to the missing cortical region, presumably there would be no cortical representation of that stimulus. Nevertheless, subjects could still point to that location, even though none of them reported that they "saw" anything. The classical explanation of this hypothetical *blind sight* was that this kind of localization behavior without conventional visual experience was mediated by a separate channel through the superior colliculus.

Recently, however, the whole phenomenon of blind sight has been challenged by Fendrich et al. (1992), who suggest that the spatial orientation phenomenon in patients with visual cortex lesions is better explained by residual vision in the primary geniculo-striate pathway. The residual vision was concentrated in very small areas of the visual field that could be detected only by very precise perimetric measurements made with stabilized visual images. The small areas of intact vision subtended only 1 degree of visual angle. Curiously, though the subjects could detect and discriminate objects, they did not report that they "saw" anything.

There may be other channels as well. Reports of a pathway to the pineal body for the control of light–dark cycles and one mediated by the superoptic nucleus controlling daily rhythms have also been reported. The point for us is that there is ample evidence for multiple pathways or channels from the eye to the brain.

2.3 THE NODES AND CENTERS OF THE CENTRAL VISUAL NERVOUS SYSTEM

So far we have spoken mainly about channels in the peripheral portions of the ascending visual pathway. We are now at a point at which we can consider the possibility that similar support for premise I may be found in the cerebral cortex itself. This expansive cap of neural tissue arching over the top of the lower centers is a wonderfully complex aggregate of a large number of interacting centers and nuclei. Visual information enters the brain mainly at the occipital or striate cortex region and is distributed from there to many other regions. The functional nature of these regions was first suggested by traumatic injuries and then by controlled studies using both microelectrode and macroscopic surgical techniques. The main idea now permeating much of our thinking about the organization of the brain is that the brain is made up of innumerable local regions, each with its own specific visual function.

The works of Allman (1981), Kaas (1978), Van Essen (1985), and, for an excellent summary, Zeki (1993) document the organizational complexity of the many areas of the primate visual brain. Allman (1981) notes that more than 14 visual areas have already been discovered in the brain of the owl monkey; Van Essen (1985) provided a map of the macaque cortex suggesting that as many as 40 heavily interconnected regions may have some visual function in that animal. Current best estimates (Felleman and Van Essen, 1991) suggest that there are, in fact, only about 30 cortical regions subserving vision in the macaque. Although the human brain has not yet been mapped in as detailed a manner, if the same kind of structural organization holds as in simpler primates, there is room for "literally dozens of visual areas in the human brain" (Van Essen, 1985, p. 283). Whatever the final number agreed on, it is surprisingly large.

The situation is made even more complicated by the intricate series of interconnections among these many visual regions. Felleman and Van Essen (1991) report that there are over 300 interconnecting pathways among these visual areas in the macaque that have been reported by neuroanatomists working on visual brain anatomy.

The conceptual model of a simple serial sequence from one kind of neuron to another that exists in the periphery is much less likely to be realized in the brain. Many pathways and, thus, many channels exist; many centers must function in an integrated manner for the various processes we collectively call "vision" to occur. Indeed, the interaction and integration of the many visual areas of the cerebral cortex is one of the most active arenas of contemporary visual neuroscience.

The problem is that it is difficult to go beyond the initial idea that multiple regions and pathways and, thus, channels exist. One source of difficulty arises from the fact that it is very hard to define precisely the function of a node in a complicated system such as the brain. Interactions of a wide variety of kinds, including convergence, divergence, lateral inhibitory and excitatory interactions, feedback, and feed-forward, all contribute to the mystery that surrounds brain function. Some suggested functions are both inventive and fantastic. In other instances, serious conceptual confusions exist among visual neuroscientists. For example, how visual information is transmitted is sometimes confused with how information is transformed into the specific neural patterns—the psychoneural equivalents—that *are* the mental experiences.

The question one must ask in the context of this very complicated system is: Is it plausible to carry out an analysis of a system of this complexity (30 or 40 areas interconnected by 300 or so tracts)? Let's simplify this question by asking only: What is the hierarchical organization of the identified visual regions? That is, what is the ordering of the arrival of signals in the various centers?

One way to approach this question is to assume, as a first approximation, that the interconnecting pathways are simple, that is, to consider them to be homogeneous communication channels conveying a collective positive or negative valence in the manner of electrostatic charge or gravitational force, and only downward, upward, or laterally. Realistically, however, it must be remembered that the system of visually involved nuclei is not so simply organized; the areas that are connected communicate by means of reciprocal, multichannel tracts consisting of many complexly encoded independent fibers. However, the proposed simplification allows us initially to consider this problem in a manner analogous to the multibody problems of mechanics.

Hilgetag et al. (1996) have shown that it is impossible to determine the hierarchical organization of the many parts of the visual brain from currently available information. They proved that there is an enormous number of equally likely hierarchies possible given the limited and idealized classification (i.e., lateral, upward, and downward) of the connection types. Furthermore, as just noted, this simple trichotomy of homogeneous connection types is certainly an oversimplification, given the variety of neuronal fiber types and functions within the intercenter pathways.

If Hilgetag and his colleagues are correct in their analysis and in their conclusions, the hierarchical arrangement of the various visual centers responsive to motion would be indeterminate from the available data. Even more important, as they go on to say, "further data, if classified by the presently understood criteria, would still not specify the exact ordering of cortical stations in the visual system" (Hilgetag et al., 1996, p. 777). To sum up: This work tells us that even the basic arrangement of the mechanisms of the brain that many would seek to correlate with perceptual data is, itself, indeterminate.

Obviously, it will be many years before the functions of even the earliest regions of the cerebral cortex are unraveled and adequately described. The complexities of this field are immense. In spite of some illusory progress using new brain-scanning tools, micro- and macroelectrode technologies, and neuropsychological case histories, we really have only the barest glimmerings of insight into the functions of these cerebral areas and know even less about how they interact. Rather than reporting what is clearly premature speculation, lets simply conclude this section by noting that the various centers do have demonstrably different functions and can be conceptualized as another example of the idea of independent nodes and channels in a complex network.

2.4 PSYCHOPHYSICAL FOUNDATIONS OF A VISION SYSTEM: INTERACTIONS AND COMBINATIONS

The realization that there are channels in the visual system that could be identified both physiologically and anatomically led to a major revolution in psychophysical research. Much of this new work has been carried out under the assumption that the activity of the channels could be independently assayed by appropriate psychophysical evidence. The evidence had always been there, even if the then-existing zeitgeist was not prepared to appreciate the significance.

Experimental designs studying the interaction of the attributes we defined earlier (color, intensity, form, texture, motion, stereoscopic disparity, perspective, shading, etc.) have proliferated in recent years. We consider this to be a promising, though still problematic, sign that vision science has now come to a mature stage in which it is appreciated that what we see is determined by many different attributes of the stimulus, not just the simple physical parameters that had been the stimuli used in an earlier kind of psychophysics.

There is also a long history of psychophysical experiments that purport to indicate that many of the modules of visual perception are independent of each other. The list is enormous, but some of the most significant that suggest that modules process the various attributes of a stimulus are: (1) Julesz's (1971) use of the random dot stereogram to show that disparity alone could produce strong feelings of depth; (2) Livingstone and Hubel's (1987) demonstration of isoluminance effects; (3) Cavanagh and Favreau's (1985) demonstration of the interaction of motion and luminance; and (4) the interaction of color and motion as reported by Gegenfurter and Hawken (1995).

In spite of this evidence, solid psychophysical data though it may be, there are formidable barriers to building bridges from psychophysics to internal neural mechanisms. It is our opinion, as expressed in another book (Uttal, 1998), that psychophysical evidence is actually neutral with regard to the underlying mechanisms. Therefore, in contrast to the usual way in which psychophysical data are presented, we do not offer this type of finding as a proof in any sense of the word that a particular kind of anatomy or physiological mechanism exists in the human visual system. At best, psychophysical evidence can only be suggestive by demonstrating that the separate attributes of a stimulus scene, if carefully manipulated by the experimenter, do interact in systematic ways.

What we do know is that some neurons do respond to specific attributes of a stimulus and are insensitive to others. We also know that some stimuli seem to be so different from others that they "pop out" from a field of other stimuli that differ from them in what are often surprisingly subtle ways. Furthermore, some stimulus attributes interact with others but other combinations seem to be independent of each other. We also know that many ambiguous stimuli can be solidified by providing additional information that constrains possible perceptual responses. We know that even such visual primitives as the perceived chromaticity of the stimulus can be affected by its surround.

The collective point of all of these observations is that what we see is multidimensionally determined. Multidimensionality, according to one argument, is the functional result of separate anatomical and physiological channels. Most important of all, however, is the repeated demonstration that as one adds attribute information to a stimulus scene, the quality of the perception improves.

It is our contention that psychophysical responses cannot define the internal arrangement of the channels. There are many important questions that cannot, in principle, be answered by any conceivable psychophysical experiment. Which particular channels? What particular mechanisms? At what level(s) of the ascending pathways does an identifiable channel exist? What is the neural language or code used in any one of the channels at any level of the ascending visual pathway? And, perhaps most important, at what level does neural activity become equivalent to the perceptual experience? These are but a few of the questions that are generated when one attempts to build bridges between psychophysics and the neural substrate, some of which cannot be answered by a behavioral methodology. To reiterate, psychophysical data are on their own terms, not reducible to neural explanations. Like mathematics, psychophysics is neutral with regard to internal mechanisms.

The reasons for this psychophysical irreducibility and neutrality are matters of both deep principle and practical complexity. In principle, one of the most fundamental reasons that we cannot go from psychophysics to underlying mechanisms can be found in the literature of automata theory. Moore (1956) proved the theorem that there were always many more possible internal mechanisms that could explain some transformation between the input and the output of a closed system than there were possible experiments. This theorem has long been known to engineers as the "black box" restriction. Without taking a black box apart, there is no way that any number of input–output measurements can tell you what machinery inside is doing; there is always a larger number of possibilities remaining than there are experiments to resolve the issue. This is an example of what we mean by an "in principle" barrier.

Other problems for psychophysical reductionism are created by the complexity of the system and the innumerable number of alternative internal possibilities. Although these are not in-principle barriers and there is no need to invoke infinite numbers as an excuse, the number of alternatives is simply so large that it would be impractical to carry out all of the necessary experiments, even if there were some brute-force approach possible.

Thus, we have concluded that psychophysics cannot independently say anything about internal mechanisms, whether they be neurophysiological or anatomical entities or even proposed component cognitive process. What we can say is that psychophysics is able to demonstrate that there are systematic interactions between experimenter-defined attributes that indicate that a set of quasi-independent processes are being executed when we perceive (i.e., consciously respond) to visual stimuli.

The word *independent* in this context has to be considered in a functional sense rather than as anything approaching concrete neuroanatomical reality. Given that there is no way to determine from psychophysical data alone the underlying neural mechanism or the level at which any process occurs, we are limited to asserting that the particular experimental paradigm being used is assaying some function of the entire system. Psychophysical phenomena cannot be localized, with the exception of a very few special situations (such as those involving the optic chiasm and dichoptic stimulation), to any particular neural apparatus.

What, then, can we ask of psychophysics? Psychophysical data can provide indirect and suggestive support for hypotheses generated either from physiological or anatomical findings or from theories and models. Thus, for example, psychophysical trichromacy suggested that there was something in the nervous system that initially established a trivariant system. Psychophysics, therefore, was the source of the idea that there were three kinds of receptor chemicals in the color system. This kind of observation, however, is neutral with regard to the specific nature of the chemicals or of the trivariant internal code or mechanisms. It was for this reason that the controversy between the Young–Helmholtz and Hering schools continued for so many years.

Among the most effective ways that psychophysical data can be used in supporting neural hypotheses is to show that there are interactions between or combinations of different stimulus attributes. That is, how, for example, do color or luminance and motion interact with motion? Do they do it in a way that is consistent with the known physiological facts? Do we find that certain combinations of attributes seem to be linked together in a way not inconsistent with what is known of the visual neurophysiology and anatomy?

Perhaps the most influential expression of how perceptual data can be used to support neurophysiological findings is to be found in the enormously influential article written by Livingstone and Hubel (1988). If there was any milestone that marked the beginning of our new understanding of channels and nodes, it had to be this one. They reviewed what was known about the anatomy and the physiology of the visual pathways. Their conclusions effectively summarize the arguments presented in earlier parts of this chapter:

> At early levels, where there are two major subdivisions, the cells in these two subdivisions [the magno- and parvocellular systems] exhibit at least four basic differences—color selectivity, speed, acuity, and contrast sensitivity. At higher stages, the continuations of these pathways are selective for quite different aspects of vision (form, color, movement and stereopsis) (p. 744).

Livingstone and Hubel (1988) go on to assert that many perceptual phenomena are associated with these magno- and parvocellular systems. The major psychophysical distinction they draw is that the magnocellular system is color-blind but has high temporal resolution and speed of response, although the parvocellular system conveys color information, but with a much slower time constant. Livingstone and Hubel offer the following key conjectures based on these anatomical and physiological distinctions:

a. Color and brightness perception have different time constants.
b. Motion perception is linked to the magnocellular system.
c. Degraded stimuli deprived of brightness information by the use of isoluminant, but colored, stimuli produce a reduced sense of two- and three-dimensional form as well as movement.

According to Livingstone and Hubel, these predictions were supported by experiments that showed the following:

The perception of movement is reduced for isoluminant stimuli (Cavanagh et al., 1984). That is, color information adds to movement perception as if it came from another channel. Therefore, there is little color selectivity in the channel mediating movement.

The perception of movement is reduced for high spatial frequencies (Campbell and Maffei, 1981). That is, spatial resolution adds to movement perception as if it was from another channel.

Movement perception is good with low-contrast images (Livingstone and Hubel, 1987). That is, lowering the contrast of a stimulus does not degrade the perception of that image. This finding implies that contrast sensitivity and motion are mediated by the same channel.

All of these results are in accord with the predicted properties of the magnocellular system. Namely, that it is a channel sensitive to contrast and motion but insensitive to color and spatial pattern.

On the other hand, the properties of the parvocellular system are not so easily linked to particular experiments by Livingstone and Hubel, but rather to some of the classic illusions of spatial arrangement. Shape- and orientation-discrimination experiments seem to show that removing color does not detract from shape and orientation discrimination. Data of this kind suggested to them that these properties were compatible with the properties of the parvocellular system.

Despite these assertions by Livingstone and Hubel (1988) and others who have followed their lead that similarities between neural and psychophysical tests are available in abundance, we believe that it is problematic whether or not these molar functional (i.e., psychophysical) similarities can be uniquely and solidly linked with any of the neuroanatomical channels. The robustness of the individual psychophysical findings, on the one hand, and of the physiological data, on the other, is indisputable. However, any specific linkages made between the two domains are at best suggestive and at worst spurious. We believe that virtually all of the visual illusions mentioned by Livingstone and Hubel in their important and influential paper are mediated by mechanisms that are far beyond the peripheral encoding mechanisms they invoke. The suggestion that such complex illusions as the attention-dependent Rubin face are the result of the different properties of the lateral geniculate body is not convincing.

What, then, can we get from psychophysical data that speaks to the two basic premises of our work? Premise I states that stimulus attribute information is analyzed into its components. Premise II states that this information is later combined or bound into a unified perceptual experience. Clearly, the basic unitary nature of our perceptual experience is strong evidence for the recombination idea expressed by premise II. Visual scenes are seen as ensembles, complete with color, movement, contrast, and pattern. If the physiological evidence is correct, then what had been initially separate is somehow joined to produce the composite experience. However, premise I has to depend mainly on the anatomical and physiological observations and findings. Unfortunately, most of the psychophysical evidence that has been invoked to support the existence of channels is of questionable value, precisely because it represents measures of overall system performance. Ascribing its properties to those of any particular level of the afferent pathway will always be an uncertain procedure at best. Our argument is that premise I cannot be supported by psychophysical evidence.

On the other hand, premise II in its general form (combination or binding exists!) can be generally supported by the nature of integrated visual percepts. However, psychophysics is as incapable of determining the exact nature of the binding mechanism as it is of specifying the exact nature of the channels.

This same point is made explicitly in an important paper by Stoner and Albright (1993). They point out that the strong interactions that occur between the various components or attributes of a visual scene make it difficult to study any one of them psychophysically in functional isolation from any of the others. For example, they review the literature and show that motion perception is dependent on how the image is segmented. Because perception is studied as a unitary whole when psychophysical methodologies are used, the dream of studying the components in isolation is precisely that—a dream.

This does not mean that the specialized components do not exist in the nervous system or that the neurophysiological coding of the separate pathways cannot be examined. Functional isolation is possible because the neurophysiological approach is fundamentally analytic rather than synthetic. This is at once both the great advantage and the great disadvantage of neurophysiology. It allows us to pinpoint at a microscopic level the workings of the nervous system and, thereby, to isolate individual functions and structures. However, it also prevents us from integrating this microscopic information into molar behavior, because this method is inherently microscopic and analytic. On the other hand, the synthetic ability of psychophysics is so powerful that it prevents any analysis. It is not clear if there is a middle ground.

2.5 SUMMARY

The purpose of this chapter has been to introduce some of the empirical physiological and psychophysical evidence that suggests that the notion of quasi-independent channels is a valid one for a model of the visual system. The concept of the visual system model that emerges is one in which multiple functions are carried out, both in sequence and simultaneously, as information is processed by a visual system. Our goal has been to show that the organizing theme of channels and nodes, separately conveying different kinds of information, which ultimately must be recombined, is supported by a substantial body of physiological fact and receives suggestive, if problematical, support from the body of psychophysical fact.

This chapter thus provides a foundation for the conceptual model that channels exist, that they segregate information on the basis of the attributes of the stimulus scene, that this segregated information is often processed at different centers in the nervous system, but that ultimately this segregated information must be recombined. It is, therefore, an anchor to biological reality that makes the modeling work we have carried out transcend engineering applications and makes this work descriptive, if not explanatory, of one of the most important sensory modalities of human existence.

In Chapter 3 we turn to one of the most challenging questions of modern science: How can the information from the individual channels and attributes be combined or bound into a unified perceptual experience? There are many answers to this question, some speculative and some based more solidly on empirical data. Whichever type one considers, they represent some of the most significant ideas of biological theory in contemporary science. The next chapter is prelude to the theoretical contributions presented in subsequent chapters.

NOTES

1. This chapter is intended to be a brief tutorial and introduction to some of the physiological and psychophysical findings of vision science for computer scientists. It emphasizes the data that speak to the issues of channels and binding, but it is not intended to be complete. Readers who like much more complete discussions of vision science are directed to such works as Wandell (1995) and Uttal (1981).
2. It is interesting to note that although a topologically constant retinotopic mapping does usually occur, it is not, in principle, necessary. The mapping could equally well be terribly scrambled (as it is, in fact, in Siamese cats, albino tigers, and quite probably to some degree in esophoric (cross-eyed) humans) and yet still be perfectly usable. All that would be necessary for such a scrambled brain to function normally is that the necessary rules of spatial relationships be present, whatever they are. It is probably for this reason that children who have had their esophoria surgically corrected for cosmetic reasons may lose stereoscopic vision. The old rules defining where corresponding points occur may no longer be appropriate for the new convergence conditions established by the surgery.

3

Models of Combination and Binding[1]

3.1 INTRODUCTION

In the previous two chapters, we described some of the ideas that have come from neurophysiology and psychophysics that have stimulated the notion of combination as a central theme of a modern theory of vision. The recent research from these psychobiological fields in which the modularity and combination ideas have become explicit, however, reflects only partially the long history of the idea of segregated modules in physiological psychology. The concept of localization, radical or conservative, has been a central theme throughout the history of this science. From the 19th century phrenologists Gall and Spurzheim to their contemporaries who were students of cranial war wounds, such as Fritsch and Hitzig, localization of specific functions in circumscribed regions of the brain has been repeatedly proposed. Localization of function (as opposed to widely distributed functions) is also a ubiquitous premise of the most modern work using brain scans such as positron emission tomography (PET) and functional magnetic resonance imaging (fMRI). In short, localized modularity has been with us for many years and continues to be a mainstay of modern brain theory.

These ideas from biology and psychology have had a substantial impact on how computational models have developed over the last few years. It is clear, however, that they were not solely responsible for the revolutionary developments that occurred in computer science. There is a strong and independent tradition of combination and modular programming among computer scientists. The late David Marr (1982) is frequently cited as the one who first proposed that separate modules function in a collaborative manner to produce the final output in vision systems. His was an idea that depended on the development of new computer technologies and adequate computer power to test theories as complex as the one he proposed. However, the ideas of modules and combination implicit in much of the early work in computer analysis of images are actually much older. Traditionally, computational image processors have worked on one attribute at a time. Techniques for finding the edges of an image, for determining the effect of a convolving filter, or for determining the shape of a surface from its shading or perspective have almost always been carried out in isolation from examination of other attributes of the image. All good programmers knew that modular programming, previously known as subprograms or procedures, and only recently characterized as "object-oriented" programs, was the way to keep one's programs from becoming an unanalyzable mess.

Marr, nevertheless, made especially important contributions to the development of the modular idea by pointing out that because some of the problems in vision were ill-posed (i.e., insufficient information was available for their exact determinist solution), a computer vision system (as well as a real organic one) could solve them only by introducing some other information, beyond that given by a specific attribute, into the computational or neural process. This other information could in many cases be in the form of additional knowledge supplied by another module sensitive to some other attribute of the stimulus image or by the assumption of some constraints, such as continuity and image smoothness, or lighting direction, or even some other information that was not actually present in the image but was part of a more extensive and previously stored knowledge base.

There are, therefore, two main streams of influence supporting the two premises that have guided us so far, one emanating from the psychobiological side and one emanating form the computer vision side. It is the purpose of this chapter to review and preview some examples of these two streams of modeling. First, we consider some of the models that have come to us from the brain and the psychological sciences. Second, we consider some of the formal models that have come from the computational or mathematical sciences. In the latter case, we also extend our discussion to consider some promising combination methods that, although they have not yet been applied specifically to visual problems, seem to be likely candidates to help us formulate and then understand this important aspect of visual perception.

Formal models and theories have been with us for some time. Mathematics and statistics have been used to describe cognitive processes throughout the history of experimental psychology. Historically, the first theories in this science, those of Weber and Fechner, were formulated in the language of algebra and calculus. Computer models have been with us from virtually the first moment that digital computer technology became available (see, e.g., Farley and Clark, 1954; Rosenblatt, 1962; Selfridge, 1958; and Widrow, 1962, among others; and for a comprehensive review of the history of this field and the mathematical relationships among the various theories, see Grossberg, 1988a; and Carpenter and Grossberg, 1991). The Anderson and Rosenfeld (1988) and Anderson et al. (1990) compendia of significant papers in neurocomputing are also good reviews of the history of one cluster of formal models. However, the recent impetus for the renewed interest in computer simulations and mathematical models of cognitive processes has been generated by, among others, Anderson (1968), Grossberg (e.g., 1968, 1969, and as summarized in his 1982 and 1988b books), Marr (1982), Rumelhart et al. (1986), and McClelland, Rumelhart, and the PDP Research group (1986). The past decade, in particular, has seen an explosion of interest in the development of formal models of perceptual processes (see e.g., Watson, 1993; Landy and Movshon, 1991).

3.2 NEUROSCIENTIFIC COMBINATION MODELS

3.2.1 The Marr Model[2]

The influence of Marr's (1982) book, simply entitled *Vision*, was enormous. It is not easy to classify this book as belonging either to the computer field or to the neuroscientific one—it had a substantial impact in both domains. Nevertheless, it is obvious that his goal was to model the processes of organic vision; for this reason we place it in this section rather than the next. Marr's ideas, it is important to appreciate, transcended the specifics of both the computer program and the psychobiology of the organic vision system by concentrating on the information-processing aspects of the problem. In the case of the kinds of problems on which he and his followers have worked, the term *processing* is very precisely defined. Marr was trying to determine what logical or computational transformations "must" be carried out to enable the perceiver to transform the properties of the stimulus pattern into those of the psychological experience. To the extent that these information transformations could be identified, then, the visual process was "understood" in Marr's view. This understanding might be obtainable even though nothing was known about the specific neural mechanisms or the (quite possibly very different) computational algorithms actually carried out in the nervous system. In Marr's (1982) words:

> The message was plain. There must exist an additional level of understanding at which the character of the information-processing tasks carried out during perception are analyzed and understood in a way that is independent of the particular mechanisms and structures that implement them in our heads. This

> was what was missing—the analysis of the problem as an information-processing task. Such analysis does not usurp an understanding at the other levels—of neurons or of computer programs—but it is a necessary complement to them, since without it there can be no real understanding of the function of all those neurons. (p. 19)

Marr then notes that there are three levels of this task with which the computational theorists must be concerned. First, the theorist must be concerned with the logic and goals of the transformation from input to output. It is necessary to determine what it is that the program must do and what strategies are suitable, if not biologically relevant, for achieving these intended goals. This level of analysis is a conceptual one and involves neither the technical details of the procedures or algorithms of the computer nor the neural logic of the brain.

Detailing the program is the essence of the next level. The theorist must then, according to Marr, commit himself to specifying the exact information processes to be simulated and the ways in which the images will be encoded or represented in order to be so processed. The specific algorithms necessary to carry out the transformation from the raw input to the processed output must be established.

Finally, at the third level, Marr argued, the theorist must be concerned about the ways in which these algorithms can be implemented physically. In fact, this third level has to be divided into two separate sublevels. First, one should be concerned with the physical implementation of the information-processing algorithms as theoretical tools. What kind of computational mechanisms could conceivably execute these processes and make the necessary transformations? This is a sublevel that is certainly achievable. However, the second sublevel of this third level of Marr's analysis implies that one must also attempt to define the actual physiological implementation in the nervous system of these processes, and it is this kind of analysis that, as we have already noted, may not be achievable.

What Marr was seeking, we believe, was a statement of the information-transforming processes that had to be executed to transform stimuli to percepts on the one hand and some speculation concerning plausible mechanisms on the other.

It is in the specifications of the information processes that "must" be executed, as opposed to either elegant computation algorithms or speculations concerning the specific neural mechanisms, that Marr's (1982) work makes its most significant contribution. In particular, he and his colleagues made great strides in understanding the nature of the computations that "must" be carried out to derive three-dimensional shape information from the two-dimensional projective images formed on the retina. Marr suggests that there are four sequential processing steps required to implement this transformation fully. The four levels of processing are:

1. The representation
2. The primal sketch
3. The 21/2-D sketch
4. The 3-D model representation

Table 3.1 (from Marr, 1982) describes the purpose of each of these steps and the primitives (or parameters of the representation) at each stage that must be manipulated to accomplish the task at each level.

Marr's work obviously ranged over a very wide prospect. In terms of general approach, it was concerned mainly with how a system might implement the transformations that it must go through to arrive at a stage at which recognition or classification of the solid objects derived from the 2-D representation can be made. Since the publication of his book we have learned an enormous amount about both the visual nervous system and psychophysics. Many of the specific details of Marr's model are no longer considered tenable, but he did set the stage for a substantial amount of modeling, much of which dealt more specifically with the concept of combination. No discussion of contemporary visual modeling would be complete without a consideration of the contribution he made to this field.

3.2.2 A Spatial-Frequency-Based Form of Perception Model

One of the most popular theoretical models of the visual system is based on the idea that a system of spatial frequency filters exists in the visual system. This idea emerged in the 1960s in the work of Kabrisky (1966) and Campbell and

Table 3.1 The Marr Theory of the Transformation Steps Involved in Stereoscopic Depth Perception (Adapted from Marr, 1982)

Name	Purpose	Primitives
Image(s)	Represents intensity.	Intensity value at each point in the image.
Primal sketch	Makes explicit important information about the two-dimensional image.	Zero-crossings Blobs
2½-D sketch	Makes explicit the orientation and rough depth of the visible sufaces.	Local surface orientation (the "needles" primitives)
3-D model representation	Describes shapes and their spatial organization in an object-centered coordinate frame.	3-D models arranged hierarchically.

Robson (1968). The basic idea embodied in what has now come to be called the "Fourier model" is a classic example of the multiple-module type of theory, in that it postulates the existence of a system of two-dimensional "filters," or feature detectors, sensitive to the individual spatial frequency components of a visual stimulus. The model hypothesizes that the visual system has a number of channels or modules that independently respond to these spatial frequencies in a way that analyzes the two-dimensional retinal image into the responses of these filters.

This physiological theory, of course, is a direct transformation of the well-known Fourier theorem in mathematical analysis. Fourier theory has been interpreted to mean that any image could be analyzed into a set of components, subject only to a very few mathematical restrictions. Originally the set of components was an orthogonal set of sinusoids, but since Fourier's time the process has been generalized to utilize, in addition, square waves, wavelets, Gabor patches, Hermite functions, and even Gaussian curves as the set of primitives. Complete discussions of the Fourier model can be found in Graham (1989) and in DeValois and DeValois (1988).

When Campbell and Robson (1968) discovered that there was selective psychophysical adaptation of a narrowly sensitive spatial frequency component, it set off what many consider to have been the mainstream of visual model building in the last two decades: the Fourier, or spatial frequency, theory of visual form perception. One of the clearest expositions of a Fourier-type model was presented by Wilson and Gelb (1984) and further developed by Wilson et al. (1990).

The premises and assumptions of the model are presented in Fig. 3.1. A stimulus is first analyzed by a set of spatial frequency filters that are selectively sensitive to a particular band of the total spatial spectrum of the image and to particular orientations. This process is formally represented in Wilson et al.'s (1990) model by the following expression. The output of one of these filters $R_{\omega\theta}(x,y)$ at a particular orientation θ and with a selective sensitivity to a spatial frequency of ω is defined by the following convolution integral:

$$R_{\omega\theta}(x,y) = \int_{-x}^{x} \int_{-y}^{y} RF(x-x',y-y')P(x',y')dx'\,dy'$$

where $RF(x,y)$ is the filter sensitivity and $P(x,y)$ is the intensity of the pattern at each point x,y.

The output of each of these spatial frequency and orientation filters is then modified by a nonlinear contrast operator, the effect of which is to compress its intensity function. This compression function is inserted into the model to match neurophysiological data that showed that the relation between a stimulus inten-

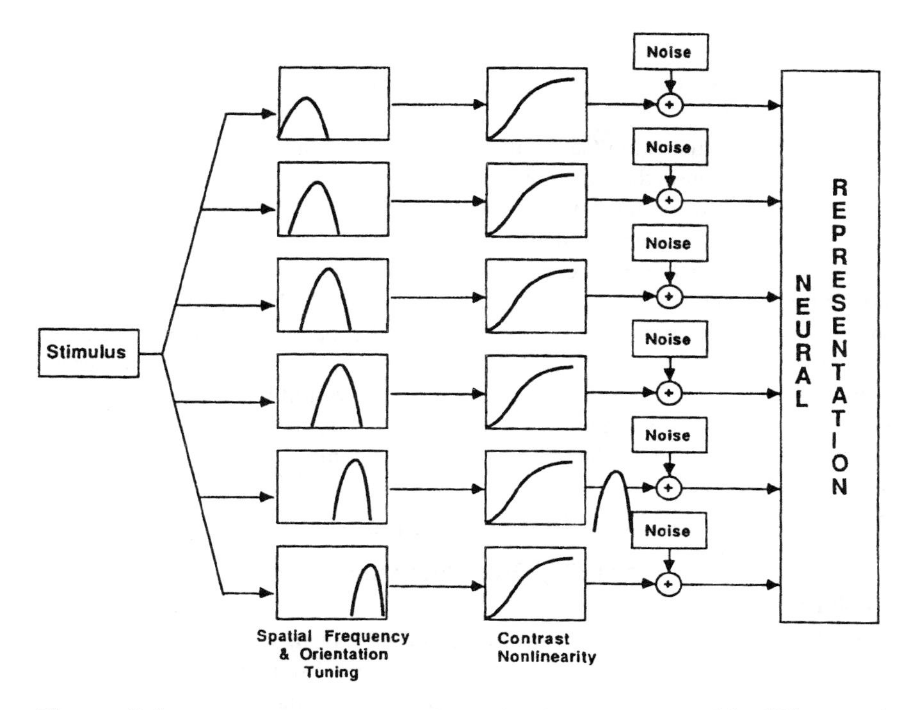

Figure 3.1 Diagram of the model of the visual system proposed by Wilson et al. (1990). (From Wilson et al., 1990. Used with the permission of Academic Press.)

sity pattern and the correlated neural response is best described by a log function or a power function with a exponent much less than 1. Some random noise is then added to each of the compressed and spatial frequency selected responses from the filter. Again, the addition of this "noise" stage of processing was influenced by neurophysiological observations that show the intrinsic variability of all neuronal components of the visual system.

Wilson and Gelb's model then goes on to develop an implicit version of the combination portion of the general model we are proposing here. The output of this set of noisy, compressed, spatial frequency and orientation filtered input pattern then becomes the representation of any input pattern *P*. Two patterns, P_1 and P_2, can then be compared by examining the difference Δ_R between the two sets of responses generated by each as expressed in the following equation:

$$\Delta_R = \sqrt{\sum_{\omega=1}^{6} \sum_{\theta=0^\circ}^{180^\circ} \sum_{-\Delta x}^{\Delta x} |R_{\omega\theta}(P_1) - R_{\omega\theta}(P_2)|^2}$$

That is, the summations are carried out for the six filters, all of the orientations that are possible, and the neighboring filters, out to an arbitrary distance designated by $\pm\Delta x$.

We refer to Wilson and Gelb's model as implicit with regard to the combination process because they do not deal with the combination process directly. The unified output is understood in their model as the vector representing the output of all of the filters as modified by the subsequent operators. In fact, this is not an unreasonable way to conceptualize the combination process. There is no a priori reason why the combination has to be effected in a particular location in the brain or to be represented by some final processing step. Whatever are the components of such a system in the brain, it is as likely that they are distributed in space as localized in a single center. This would be entirely compatible with the notion of population statistics (as opposed to a pontifical individual neuron) that may well be the way the brain is organized.

3.2.3 Graphical Combination Models

Bülthoff and Mallot's (1990) Graphical Model of Depth Perception

A number of psychobiologists have modeled data from various experiments in a much less formal manner than characterizes the Wilson and Gelb formalisms. A block diagram is often used as the medium for this kind of attempt to summarize a variety of empirical data. One example of this was presented by Bülthoff and Mallot (1990) in their summation of data describing how stereo, shading, and texture are combined. The psychophysical evidence on which the model was based is a series of experiments in which the interactive effects of edge-based stereo disparity, shading-based stereo disparity, and shading itself were examined. Using the perceived elongation of a stimulus object as an estimate of the perception of depth, Bülthoff and Mallot determined that this experience was enhanced when additional information was provided by any of the three cues. They found a nearly perfect fit between perceived depth and the actual depth of the displayed stimulus when all three of these cues were provided. However, they also observed a decline in this fit when edge information was removed. Even more surprising was that little perceived depth was reported when the shading cues were presented alone.

The most interesting parts of their work, however, were those situations in which conflicts were deliberately created among the cues. This is a useful means of determining the priority or precedence of each of the attributes cum cues in competitive situations. In these cases they found that edge-based stereo always dominated the shading and disparity cues. In fact, edge-based stereo was so powerful that it was able to veto either of these other two cues. Stereo from disparate shading, on the other hand, reduced but did not completely overwhelm the cue to monocular shape from shading.

Based on these kinds of data, Bülthoff and Mallot (1990) proposed the combination model of visual perception shown in Fig. 3.2. In this diagrammatic model, the width of the channels feeding from each of the boxes to the final perceived output metricizes the impact that each type of cue has on the final perception of depth.

Cavanagh's (1987) Graphic Model of Depth Perception

Although not dealing directly with precedence or dominance of one cue or another, another important combination model of visual depth perception has been presented by Cavanagh (1987). Cavanagh's main point was that many different attributes or cues of the stimulus scene contribute to our ability to see depth and the fine structure of surfaces. He proposed that certain of the scene attributes we discussed in Chapter 1 were combined, and thus transformed, into intermediate-level superattributes. For example, shadows and subjective contours were transformed from the raw luminance distribution pattern. As another example, form was a superattribute derived from a combination of all of the other primary attributes. These three intermediate-level attributes were then combined with

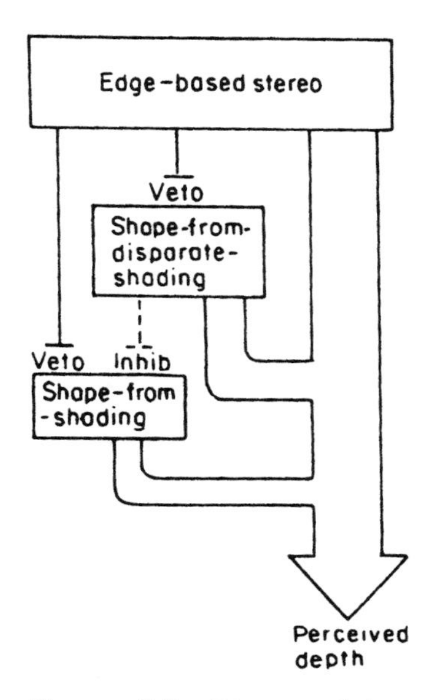

Figure 3.2 Diagram of the model of the visual system proposed by Bültoff and Mallot. (From Bültoff and Mallot, 1990. Used with the permission of Wiley, Publishers.)

motion and binocular-disparity information at some higher level where computations, inferences, and representations produced the observer's perception of the depth and fine surface structure of the original scene. Cavanagh's (1987) diagrammatic model of segregated attributes, channels and recombination is shown in Fig. 3.3.

Some Other Models

Many other theories incorporating segregation by modules sensitive to one or another aspect of the stimulus scene and subsequent combination of the information contained in these separate "channels" have been proposed. Treisman (1986a), Treisman and Gelade (1980), and Treisman and Patterson (1984) have

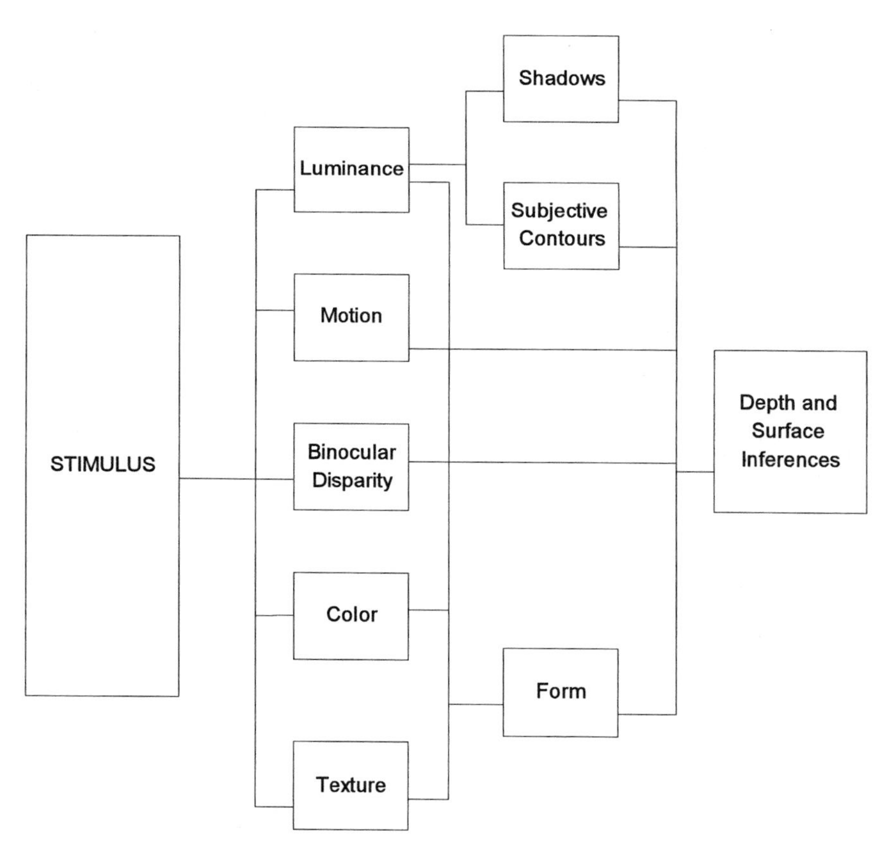

Figure 3.3 Diagram of the model of the visual system proposed by Cavanagh. (Redrawn from Cavanagh, 1987.)

all described a similar model, in which attention plays a key role in determining which features are combined.

3.2.4 Freeman's Temporal Synchronization Model of Combining

All of the models discussed so far imply that the signals from the various attributes converge on a single spatial location in the nervous system where they are combined or bound together to produce a coherent and unified perceptual experience. As mentioned elsewhere in this book, spatial convergence is not the only possible mechanism for attribute or feature binding. It is also conceivable that convergence in time could account for the binding of signals from separate channels or from what have commonly been called neural feature detectors. The long-term champion of such a temporal synchronization approach to binding has been Freeman (1975), who reported the discovery of a 40-Hz electroencephalographic signal in the olfactory system of the cat that he believed represented either a synchronizing signal that forced organized firing of all of the cells that responded to chemical stimuli or the synchronized activity itself. This cyclic electrical brain activity produced (or was), according to this theory, a phase locking of the activity induced in each special channel or feature detector by its own "trigger feature" of the stimulus. To the degree that the activity was so synchronized, it would produce a unified perceptual experience of whatever stimulus had been applied. To the degree that the signals were not synchronous, the response would be disorganized and presumably unperceived.

Unfortunately, the temporal binding hypothesis is as difficult to study as is the spatial combination one. The main tool used by Freeman was the electroencephalogram. This global measure of neural activity unfortunately integrates the activity of the individual cells in a way that obscures what the individual channels may be doing. The problem of observing what neurons in many different channels in the nervous system are doing at the same time is even more challenging than determining what they are doing at different times. It simply is not yet technically possible to determine physiologically if a reasonably sized and relevant set of neurons was actually being synchronized at the time a perceptual experience occurred.

One criticism of the Freeman model, actually raised by himself in a later work (Freeman, 1995), was that no readout mechanism was proposed that could "interpret" this widely dispersed but temporally synchronized pattern of neural activity. It seems to us, however, that this is not a real issue. Just as a spatial pattern of activity could itself be the experience, so too could a temporal pattern. Temporal patterns are certainly codes at lower levels of the sensory pathways and could well be the psychoneural equivalents of perception at higher levels without the necessity of invoking any interpreter or homunculus-like entity. The main difficulty with this approach is its empirical intractability—how one would test the idea of synchronized but dispersed codes remains uncertain.

3.3 COMPUTATIONAL COMBINATION MODELS

The models we have described so far have been built primarily on the basis of data emerging from psychophysical, neurophysiological, and anatomical findings. But there is another thrust that has come specifically from mathematics and computer science. The aim of this alternative approach has, nevertheless, been the same as the psychobiological one—to describe how information obtained from more than one source can be combined to produce a better estimate of the nature of an acquired scene than would be possible with any single-attribute database. The sources of this information may not be explicitly biological but may come from engineering experience with video sensors, from algorithms that act on sensed data, or from decision-making programs that produce highly refined or transformed versions of the acquired data. Computer scientists have developed these techniques to accomplish what are essentially the same goals for which the organic system has evolved—to suppress noise and to improve reliability. Although some of these methods have been applied to the psychobiological problems we are concerned with in this book, some that we discuss in the following section have not. Nevertheless, some of those not yet applied to the vision problem are so promising that we have chosen to include a discussion of them in this book.

The basic idea that we have championed so far remains the same, however. That idea is that powerful and robust representations can best be obtained by combining several image-processing algorithms or processes. This is another way to express both premise I and premise II presented in the biological context of Chapter 2. For example, suppose that it is desired to segment an image that consists of several two-dimensional textured regions, and we have available to us several texture segmentation algorithms. Individually, each algorithm, when applied to the image, may falsely indicate the presence of a texture boundary at least some of the time and may also miss parts of the true boundaries at least some of the time. Usually, the errors made are different for each algorithm, making it theoretically impossible to obtain a true demarcation of the texture boundaries from any one algorithm's output. In order to overcome this handicap, it is necessary to devise a technique for combining all of the algorithms' outputs in a way that cancels individual errors. We present one novel method for combining texture segmentation algorithms in Chapter 5.

Beyond two-dimensional texture segmentation, however, data fusion, or attribute combination, has been developed to meet the needs of computer vision tasks, for example:

Integrating depth information obtained from shape-from-shading with that from stereo
Recognizing an object from its texture, color, and shape
Combining range data from ultrasound and infrared sensors

The field of algorithm combination, or, as it better known in computer vision circles, data fusion, has become extremely important in the processing of visual images, as evidenced in three very important recent books (Abidi and Gonzalez, 1992; Aloimonos and Shulman, 1989; and Clark and Yuille, 1990). Data fusion ideas, however, have application far beyond the field of vision, the main topic of this book. For example, developing the technology behind the integration of radar, sonar, infrared, and other sensors is likely to be one of the most important military goals of the next century (Dasarathy, 1994).

The aim of this part of this chapter is to provide a brief introduction into a group of mathematical combination methods, some of which have already been applied to problems in vision science and others of which show promise in solving some of the vexing problems faced in this field. These include:

Bayesian statistics and Markov random fields
Dempster–Shafer theory
Fuzzy logic
Lebesgue logic

Each of these theories individually, as well as all of them collectively, has already provided or is likely to provide "languages" in which data fusion problems can be formulated and the effects of various combination rules can be studied. The language for each theory is described here; whenever possible, an example is given to illustrate how data fusion problems may be formulated in that language. When possible, a proposed or already-implemented application to vision is included with the description of each theory.

3.3.1 Bayesian Statistics

The modern axiomatic theory of probability began with the publication of A. N. Kolmogorov's (1929: 1956) *Foundations of the Theory of Probability*. In that work, Kolmogorov defined the concept of a probability measure and elucidated its significance for the rigorous study of stochastic phenomena. Probability measures are commonly used in many disciplines. However, since they play such an important role in the analysis of data fusion and combination techniques in vision and other sciences, it is worth briefly reviewing their formal definition and properties in the context of this particular discussion.

The starting point for probability theory is to define a set Ω of all possible outcomes of an experiment. Such a set of all possible outcomes is referred to as a *power set*. To illustrate the discussion that follows, let us consider, as an example, rolling a die. In that case:

$$\Omega = \{1,2,3,4,5,6\}$$

Here each number represents what the die might show when it stops rolling. We may be interested in assigning probabilities to some, but not all, subsets of Ω, but we would like to do so in a consistent way. To make this possible, we need to construct a collection of *events*: subsets of Ω that are to be given their own probabilities of occurrence. For example, the subset $\{1,3,5\}$ is the event that the die shows an odd number, and we may give this a probability of 50%, or .5. Kolmogorov postulated that any collection of events that are given probabilities must have the following structure:

a. If the collection includes an event A, then it must include also the complement of A, denoted A^c, which is the set of all possibilities not in A. Intuitively, this means that if we assign a probability to an event's occurring, then we must also be able to assign a probability to the event's not occurring.
b. If the collection includes two events A, B, then it must include their union, $A \cup B$, i.e., the logical OR. (In fact, this must be true even for the union of an infinite number of events.) This means that if we assign a probability to any two events, then we must assign a probability to the occurrence of at least one or the other. For example, suppose $A = \{1,2,3\}$, the event that the die shows less than 4, and $B = \{4\}$, the event that the die shows 4. Then $A \cup B$ is the event that the die shows less than 5, i.e., $A \cup B = \{1,2,3,4\}$.

An example of a collection of events meeting conditions (a) and (b) is the set $\{\Omega,\emptyset\}$, where $\emptyset$ is the empty set. This is a trivial collection, for we are able to assign probabilities only to the certain event and the null event. A more interesting collection of events is the set of all possible subsets of Ω, which in this case would have a total of $2^6 = 64$ possible events. This collection is called the *power set* of Ω, the name originating from the power law that describes how many members it has.

Any collection of events that satisfies conditions (a) and (b) can be assigned probabilities. Any assignment of probabilities is called a *probability measure* P, and Kolmogorov postulated that it should have these properties:

a. $P \geq 0$ and $P(\Omega) = 1$. These conditions mean that probability is a nonnegative quantity, whose maximum is 1 for the certain event.
b. For any two sets A, B, with no elements in common, $P(A \cup B) = P(A) + P(B)$. This condition requires that probability "add up" properly so that the total is exactly the sum of the probabilities assigned to the parts. A conse-

quence of this rule is that the probability of an event's not occurring is simply 1 minus the probability of occurrence:

$$P(A^c) = 1 - P(A)$$

These postulates may seem simple and innocuous, but their implications for probability and its application to the real world are profound. Many statisticians have suggested that they cannot be the sole basis for a theory of probability; other axioms are possible that can lead to alternative predictions in physics, engineering, and other fields. For an interesting discussion of the implications of the probability axioms for quantum mechanics, see Gudder (1988), and for the implications in engineering applications, see Gardner (1988).

Returning to the example of the rolling of a die, suppose that we want to assign probabilities to various outcomes. To that end, we roll the die a large number of times, say 100. A typical set of outcomes is shown in Table 3.2. From these data, a probability measure can be assigned: $P(\{1\}) = 16/100$, $P(\{2\}) = 15/100$, etc. It is reasonable to wonder if these numbers tell us that we have a fair die, because the probabilities are not exactly the same for all outcomes. Although this question can be "settled" by applying a statistical test such the Kolmogorov–Smirnov, it is worth examining another approach, the one we are discussing here—data fusion.

Suppose that we run another set of 100 trials. The outcomes may be slightly different; in particular, we may observe that this time $P(\{1\}) = 17/100$, $P(\{2\}) = 16/100$, $P(\{3\}) = 16/100$, $P(\{4\}) = 17/100$, $P(\{5\}) = 16/100$, $P(\{6\}) = 17/100$. We can combine these two observed probability measures by simply taking the average:

$$P = \tfrac{1}{2} \times P_1 + \tfrac{1}{2} \times P_2$$

where P_1 and P_2 are the probability measures observed in the first and second sets of trials, respectively. Note that the resulting function P satisfies all of the conditions to be a legitimate probability measure. Furthermore, whatever P_1 and P_1 are, the new measure P is never a less uniform probability measure. That is, it is never further in distance (as measured by the Kolmogorov–Smirnov test) than *both* P_1 and P_1 are from the uniform measure, which gives probability 1/6 to each of the die's faces. If we adopt this combination procedure and combine

Table 3.2

Die shows	1	2	3	4	5	6
Number of times	16	15	17	15	18	19

many more observed probability measures by taking the average of all, it is clear that for a "fair" die we will eventually converge to a truly uniform assignment of 1/6 to each of the six faces on the die[3].

The data fusion example shows how simple averaging is useful for combining probability measures. However, although averaging is simple, it does not give a way of discriminating between conflicting and consistent data: averaging suppresses conflict among data, even when that conflict is telling us that something is wrong. For example, suppose that our die is not really fair, and the second set of trials produces the probability measure

$$P(\{1\}) = 30/100, \qquad P(\{2\}) = P(\{3\}) = \ldots = P(\{6\}) = 14/100$$

These results suggest that the die is "loaded" to show "1." But simple averaging removes this anomaly, producing a combined probability measure that increases uniformity. This means that averaging should be used only when there is no need to check for conflicts in data.

The tendency for averaging to increase uniformity is a disadvantage, but it still can be used to combine data. In fact, in some situations it may be the simplest and most computationally efficient means. However, a more powerful method that is widely used in probability theory is based on Bayes' rule, which can be stated in the following way. Suppose that we have some data d, and we want to use it to decide whether a conclusion H is likely among a set of other conclusions. The *likelihood* of data d arising if H were true is defined to be the ratio

$$L(d \mid H) = \frac{P(d \mid H)}{P(d \mid H^c)}$$

The term in the numerator is the probability of the data d's occurring if the conclusion H is true; the denominator is the probability of d's occurring if H is not true. This ratio is a positive number with no upper limit (unlike a probability), and the higher it is, the more likely the data. A similar ratio can be used to define the a priori odds of a conclusion's being true:

$$O(H) = \frac{P(H)}{P(H^c)} = \frac{P(H)}{1 - P(H)}$$

Bayes' rule can now be stated. It says that the a posteriori odds of a conclusion H's being true after data d have been observed is the product of the likelihood with the a priori odds:

$$O(H \mid d) = L(d \mid H)O(H)$$

This rule can be applied recursively to update odds as more data become available. In particular, the a posteriori odds of H after independent data d_1 and d_2 are observed is

$$O(H \mid d_1, d_2) = L(d_2 \mid H)O(H \mid d_1)$$

To illustrate the Bayesian method of data fusion, let us return to our example with the die. Suppose that the manufacturing process for dice is such that 99 times out of a 100 the die is fair, meaning that each of six sides is equally likely, and the remaining time the die is loaded to show 1 approximately 84% of the time. To determine whether the die is fair, we roll it 100 times and observe 16 instances of 1; and on a subsequent set of 100 rolls, we observe 30. The a priori odds of a fair die are $O(\text{fair}) = 99$, based on the manufacturing statistics. The likelihood of observing 16 instances of 1 in the first set of 100 rolls is (from the binomial distribution) $L(16 \mid \text{fair}) \cong 0.10$. Therefore, the odds of a fair die if 16 instances of 1 are observed are $O(\text{fair} \mid 16) \cong 10$. Note that the a posteriori odds are reduced over the a priori odds, because simply observing 16 instances of 1 isn't enough to conclude much about the die's fairness: we would have to observe the frequencies of the other faces as well. Now the odds of a fair die after the second set of trials should be even less because of the disproportionate number of 1s observed. Indeed, calculation shows that $O(\text{fair} \mid 16{,}30) \cong 0.0045$. Thus, we can virtually rule out the conclusion that the die is fair.

Bayesian methods allow us to combine data without the uniformity effect of averaging. Although Bayes' combination rule is simple and powerful, it has the same drawback as the simple averaging rule: there is no way to check if two sets of data are so inconsistent that they should not be combined. Nevertheless, Bayesian statistics has found many applications in data fusion applications to medical diagnosis and other areas (Shafer and Pearl, 1990).

Markov Random Field Models

An important example of the use of Bayesian methods in image processing is provided by the image-restoration work of Geman and Geman (1984). There, a useful prior distribution on the space of images is introduced (thus allowing the calculation of a priori odds for images), and it is shown how the powerful optimization techniques of simulated annealing may be used to estimate the maximum a posteriori image from blurred and noisy data. Geman and Geman's methods are aimed at image restoration and don't, by themselves, accomplish data fusion; however, an extension of their work for the latter purpose has been provided by Poggio et al. (1988). Both Geman and Geman's and the Poggio group's work is now reviewed.

The starting point is general pattern theory, which provides a means for assigning probabilities on the space of images (Grenander, 1993). Consider the set containing all possible images: it is, of course, infinite, and even if we restrict ourselves to binary images of a certain size, say, 64×64 pixels, this set is still gigantic, containing 2^{4096} possible images. With such a large number of possibilities, how can one assign probabilities to certain events of interest, such as the image's having a periodic structure? These sorts of questions are the motivation behind the theory developed by Grenander, which allows the creation of abstract representations for, and assignment of probabilities to, images in terms of relationships between pixels and their neighbors.

General pattern theory starts with the simple observation that the structure of most images is such that there is some correlation between the value of a pixel and those of its immediate neighbors, but there is little if any correlation between a pixel and far away pixels. Furthermore, the correlation between neighbors changes drastically where the neighbors lie on different sides of an edge. A Markov random field (MRF) model is essentially a mathematical formulation of these facts. An MRF is a random process on the plane, for which the sample images are such that the probability of a pixel $x_{i,j}$'s having intensity w (being white, say) depends only on what the intensities are of the four connected pixels $x_{i-1,j}$, $x_{i+1,j}$, $x_{i,j-1}$, $x_{i,j+1}$ (these are, respectively, the pixels immediately to the left, right, above, and below the pixel $x_{i,j}$), and not on the intensities of pixels further away than those four. There are exceptions if an edge passes between $x_{i,j}$ and one or more of its neighbors. Strictly speaking, the preceding describes a particular type of MRF, called a *nearest-neighbor* MRF. Other types of MRFs are similar in structure, and they model correlations between pixels and larger neighborhoods greater than the one including only the nearest neighbors.

For illustrative purposes, consider as an example the synthetic image consisting of three polygonal regions (a black triangle, a black rectangle, and a hatched rectangle) superimposed on a white background, as shown in Fig. 3.4. Within each of the black polygons, and on the white background, there is a perfect correlation between the value of a pixel and that of its neighbors. This correlation changes drastically at the edges of the polygons, however. Furthermore, the correlation is weak within the hatched polygon. The pixels in the black polygons, on the other hand, have a relatively high probability of occurring with a nearest-neighbor MRF.

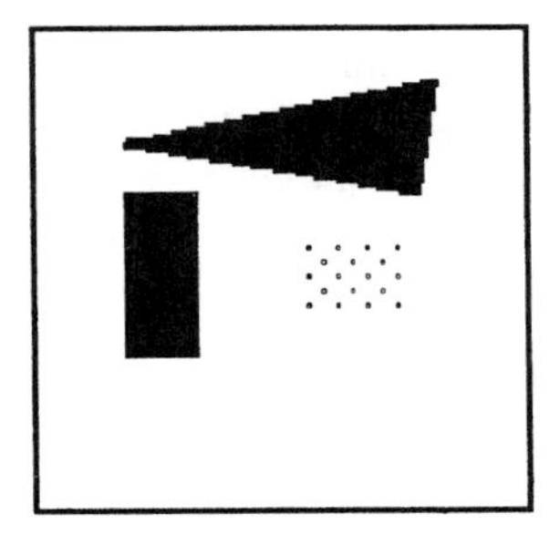

Figure 3.4 A sample visual problem posed by three objects differing in shape and texture.

From a mathematical point of view, it is worth noting that MRFs are particularly easy to work with as models for images, because they lead to an exponential form for the probability distribution of the space of images. (Distributions that have an exponential form, such as the familiar normal distribution, are among the easiest to analyze and fit to data in practice.) The exponential distribution that describes each MRF is called its *Gibbs* distribution, and it is written:

$$p(x,e) = \frac{1}{Z} e^{-U(x,e)/T}$$

Here the symbols employed have the following meanings: x is the image, represented as a matrix of pixels; e represents the edges in x; $U(x,e)$ measures the differences in intensity between neighboring pixels in x, except where edges occur; and T, Z are constants for scaling and normalization, respectively.

Suppose now that we observe a blurred, noisy version of the blocks image of Fig. 3.4, as shown in Fig. 3.5. In an image such as this, the underlying "signal," which in this case consists of the simple blocks, can be modeled by an MRF as noted earlier, giving in turn a simple Gibbs distribution for the a priori odds on the possible signal images. The "noise" in the image, which is spatially uncorrelated, can also be modeled by an MRF, specifically one where there is no dependence between a pixel and its neighbors. Noting that MRF models are possible for both signal and noise separately, Geman and Geman showed that by Bayes rule one can also construct an MRF model for the *observed* image, containing both signal and noise. The significance of this result is that a Gibbs distribution exists giving the a posteriori odds $O(x,e|g)$ for the signal image x and its edges e, given the observed image g. One can then accomplish image restoration by finding the maximum a posteriori estimates for both x and e given g, by finding choices for x and e that maximize the function $O(x,e|g)$. Geman and Geman showed that the maximization problem can be solved by using an algorithm based on the powerful technique known to optimization researchers as *simulated annealing*. This algorithm is guaranteed to con-

Figure 3.5 A noisy, blurred image of the problem presented in Fig. 3.4. Depending on the type of noise, one of the figures can completely disappear.

verge always to the true maximum, although only if infinite time is available, and then only in a weak sense of convergence known as *convergence in distribution.*

Geman and Geman's (1984) work is one of the widely cited in the field of image restoration. However, this work does not address the problem of data fusion. Toward that end, T. Poggio and coworkers (Poggio et al., 1988; Gamble et al., 1989) have introduced the idea of coupled MRFs, which extends Geman and Geman's work to provide integration of outputs from different modules in a vision system. In their extension of this work, the aim of the Poggio group was to classify the edges in a scene as arising from either a shadow, changes in albedo, changes in surface orientation, or an occluding boundary. It is possible to distinguish one type of edge from another by examining the responses of modules processing the scene on the basis of hue, texture, motion, or depth from stereo. Each type of edge may be detected by one or more of the modules. However, a classification as to physical origin requires that the edge maps in the outputs of *all* modules be examined, as no one module is sufficient to determine all origins.

Suppose that there are N modules in all. Each module produces an edge map for the scene, but these maps are not in general aligned with each other, because the placement of edges is subject to some uncertainty. It is necessary to introduce some means of alignment in order later to be able to classify the origin of the edges by examining the outputs of all N modules. Poggio and his colleagues use the edge map obtained by the brightness module as the "ground truth" in order to align all of the edge maps from the other modules. The brightness edge map, which we denote B, is used to align the edge maps produced by the other modules in the following way. The edge map to be produced by the kth module can be viewed as the maximum a posteriori estimate from a Gibbs distribution of the form:

$$p(x^k, e^k \mid g) = \frac{1}{Z} e^{-[U(x^k e^k \mid g) + V_C(e^k, B)]/T}$$

This distribution is essentially the same as introduced earlier, but it now incorporates the term $V_C(e^k, B)$ to measure the alignment between the edge map e^k for this module and the brightness edge map B. Consequently, the maximum a posteriori estimate for the edge map e^k must not only fit the data g, but also be aligned with B as far as possible.

Once all of the edge maps are aligned with B, it is possible to use a linear classifier that estimates the origin of a physical edge by pooling the responses of all modules. Suppose there are M types of edges in all. At the (i,j)th pixel, the odds given by the classifier to the event that edge type m has occurred, for $m = 1, \ldots, M$ is obtained by the linear combination:

$$\sum_{k=1}^{N} W_{mk} e_{ij}^{k}$$

Here the classification weights W_{mk} are obtained by using a neural network on a suitable training sequence (Gamble et al., 1989). The edge type with highest odds is the one chosen by the classifier.

The key to the application of MRF models to data fusion by Poggio and his colleagues is the use of an explicit representation for edges in the Gibbs distribution. The exponential form of this distribution makes it attractive for purposes of calculating the maximum a posteriori estimate. Although the algorithms involved in maximizing the Gibbs distribution require extensive time on a serial computer, there is considerable scope for parallel implementation.

We now turn to another application of Bayesian techniques and MRFs to data fusion.

Jeong and Kim's Bayesian Unification Theory

Jeong and Kim (1992) have developed the Bayesian concept into a model of early vision. The interest was specifically in the "unification" or combination of features and is based on a philosophy much like the one presented in this book. Their system first analyzes and then synthesizes incoming visual information in accord with premise I and premise II presented in Chapter 2.

Jeong and Kim's model, shown in Fig. 3.6, is probabilistic in the Bayesian sense. It uses a somewhat different set of stimulus attributes than some of the others already discussed, but the concept is generally the same. Their model identifies intensity, optical flow, disparity, shading, and texture as the critical attributes. It is a dynamic theory in which comparisons of the left and right images from two eyes or cameras are buttressed by temporal information obtained at two different times. The goal is to define the three-dimensional shape of the seen object as well as its motion.

In Fig. 3.6, we see that the two images from the left and right eye or camera (identified as f^l and f^r, respectively) are first processed to segment and refine the image. In the Jeong and Kim model, they use the term *restoration* in the same way we use the term *preprocessing* or *segmentation*. This part of their model consists of methods to define edges and to determine the best possible estimate of the intensity distribution of the images. The edges and intensity distribution are carried out by a neural network model in which local interactions define the connectivity (as a cue for intensity distributions) and enhance the edges in a manner quite similar to that performed by lateral interactions in the organic retina.

The output—the intensity and the line processes—of this restoration process are considered to be the a priori information and the basis for further manipulation. In Fig. 3.6, the information from the restoration process is defined as: g^l,

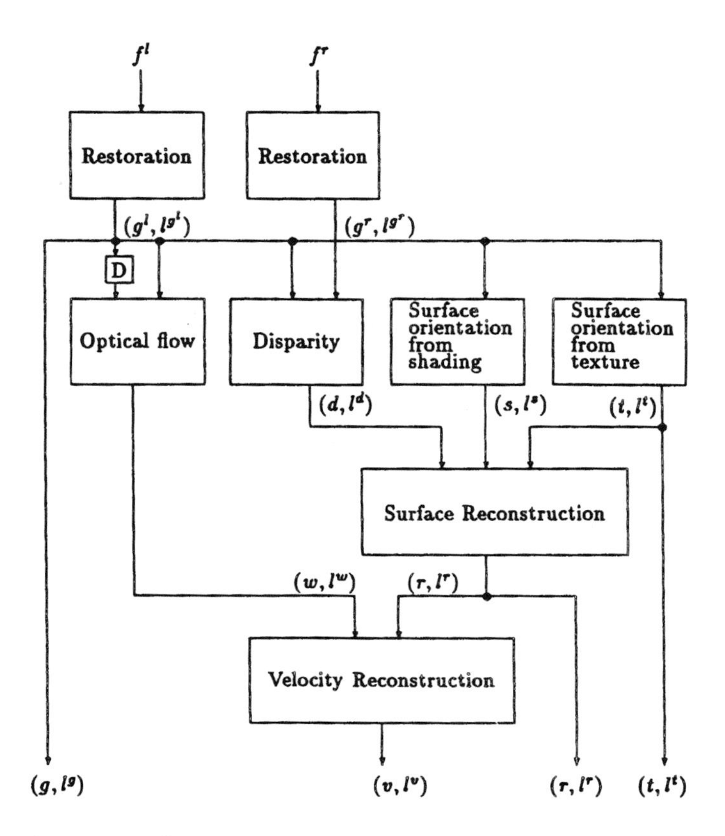

Figure 3.6 Diagram of the model of the visual system proposed by Jeong and Kim. (From Jeong and Kim, 1992. Used with the permission of the IEEE.)

which is the local intensity information from the left image; l^{g^l}, the edge information from the left image; g^r, the local intensity information from the right image; and l^{g^r}, the edge information from the right image. This is the raw information used by subsequent stages of the model.

These data, which represent the prior knowledge, is then inserted into a Markov random field formulation (Geman and Geman, 1984), as just described. In addition, two other constraints—noise disturbance and smoothness—are invoked to provide the criteria for a minimum-energy solution of the problem. Disparity is computed directly from the left- and right-hand images using this information and these two constraints to minimize the energy functional representing the disparity.

The motion or optical-flow model carries out the process in a similar way, but rather then comparing left and right images, it compares g^l and l^{g^l} taken at time 1 with the same measures taken at time 2. This is represented by the term D (delay) in Fig. 3.6.

Texture and shading are used independently to provide two estimates of surface orientation: surface orientation information being needed along with disparity to reconstruct the surface. At this point d, s, and t represent the disparity-driven, shading-driven, and texture-driven estimates of the surface, and l^d, l^s, and l^t represent the comparable edge data. The block designated "Surface Reconstruction" represents one of the two parts of the synthesis or combination portion of Jeong and Kim's (1992) model. Its inputs, in accord with the Bayesian formulation, are the prior estimates of these values, and its outputs are r and l^r, the reconstructed values of the surface and its edges.

r and l^r now become the inputs (along with w and l^w—the outputs of the optical flow operator) to the velocity reconstructor, another synthesizer or combiner based on the Bayesian and MRF ideas just discussed. Finally, all of the outputs are combined to represent collectively both the dynamic and the static properties of a moving surface.

In summary, Jeong and Kim's (1992) model has taken us from the original pair of two-dimensional input images to a full-scale reconstruction of a moving three-dimensional surface. It did so by first analyzing the images into a set of attributes and then recombining them into a full description of the scene. The mechanism used for the recombination process was Bayesian, in that certain pieces of information were used as priors and then, in accord with the MRF energy-minimization procedures, posterior estimates were produced, as constrained by noise and smoothness requirements.

3.3.2 Dempster-Shafer Theory

Probability theory gives us no way to take into account "ignorance," and so can lead to ludicrous predictions. This fact has led some statisticians to suggest other ways to approach the problems confronted in visual combination and fusion. One of these alternative approaches is known as Dempster–Shafer theory. Consider, for example, the following example, similar to that presented in Shafer (1976). Suppose that we wish to assign probabilities to the outcomes of an experiment, which uses an infrared scanner, that is designed to test whether there are "ghosts" in a room, i.e., apparitions that are similar to people who are already dead. One might group the outcomes of this experiment into three categories: (1) there are no ghosts in the room; (2) there are ghosts, but the scanner does not indicate their presence; and (3) there are ghosts, and the infrared sensor indicates their presence. If we were completely uncertain about the outcome of this experiment, then we would give a probability of 1/3 to each of the three categories. By making this assignment, we have expressed our ignorance by spreading the probabilities evenly across all possible outcomes. However, given this uniform assignment, suppose we were to calculate the probability of the event "there are ghosts." In terms of the categories just used, this event consists of categories (2) and (3). Therefore, complete uncertainty means that the probability of there being ghosts is $1/3 + 1/3 = 2/3$! It is worth noting that we reach this "conclusion" only by starting from a viewpoint that uncertainty means that all outcomes are equally probable.

In many, if not most, situations there is ignorance about the true state of affairs. It is not satisfactory, however, to model this ignorance by using a uniform assignment of probabilities computed by taking the average, as the previous example shows. To overcome this problem of being forced to make uniform assignments, an alternative theory to probability has been developed by Dempster and Shafer (Shafer, 1976). The main premise of this theory is the realization that ignorance can be formulated.

The Dempster–Shafer theory begins by constructing for each experiment a *frame of discernment* Ω. This is simply a set containing all possible outcomes, much like the universal set used in probability theory. The collection of events is taken to be the power set of Ω, which the reader may recall from an earlier section is the collection of all possible subsets of Ω. The assignment of probabilities is represented by a function m, called a *basic probability assignment* (bpa), which satisfies the following conditions:

a. $m(\emptyset) = 0$. This means that no probability is assigned to the never-occurring event.
b. The total of the probabilities assigned to all of the subsets of Ω is 1.0. This condition guarantees that the probabilities add up to unity when all possible outcomes are considered.

The bpa is different from a probability measure P as defined by Kolmogorov. In particular, we need not have $m(\Omega) = 1.0$, whereas this is required of every probability measure on Ω. In fact, the quantity $1 - m(\Omega)$ represents a measure of the ignorance that we acknowledge to be present in making this probability assignment. Furthermore, for any two disjoint events A and B, it need not be that $m(A \cup B) = m(A) + m(B)$. We could have, for example, that $m(A \cup B) < m(A) + m(B)$, or even that $m(A \cup B) > m(A) + m(B)$.

From the bpa, it is possible to determine the level of belief in any event; denoted Bel(A), this is defined to be the sum of values given by the bpa for all subsets of that event; i.e.:

$$\text{Bel}(A) = \sum_{X \subseteq A} m(A)$$

Note that by property (b) of bpa's, this means it must be that $\text{Bel}(\Omega) = 1.0$.

From the point of view of data fusion, the chief attraction of the Dempster–

Shafer theory is the Dempster combination rule. This rule provides a means of combining two bpa's m_1 and m_2, to yield a new bpa, denoted $m_1 \oplus m_2$. Since beliefs are defined in a simple way from bpa's, this means that we can combine belief functions simply by combining their underlying bpa's. For any event A, the value of the new bpa $m_1 \oplus m_2(A)$ for that event is defined to be the sum of the products $m_1(X)m_2(Y)$, where X and Y both contain A as a subset. The new bpa is then normalized so that it gives a total probability of 1.0 when added up over all subsets of Ω. More precisely, we have the equation

$$m_1 \oplus m_2(A) = \frac{\sum_{X \cap Y = A} m_1(X)m_2(Y)}{1 - \sum_{X \cap Y = \emptyset} m_1(X)m_2(Y)}$$

For this combination procedure to work, it must be true that the two bpa's have at least one set A in common to which they both give nonzero probability; otherwise they are deemed inconsistent, and no combination is possible. (In the equation just given, the denominator is zero for any inconsistent pair of bpa's.) In this important way, the Dempster combination rule differs from simply forming the average $(m_2 + m_2)/2$, which would also give us a way of combining two bpa's but without regard to their possible inconsistency.

The role of bpa's, belief functions, and the Dempster combination rule is illustrated in the following example. The presentation of this example is inspired by the review of Dempster–Shafer theory by Gordon and Shortliffe (1990). The facts about color vision that are used next are taken from the Optical Society of America's Committee on Colorimetry's (1953) encyclopedic work *The science of color*.

An Application of Dempster–Shafer Theory in Vision

Human color vision is normally trichromatic; that is, three colors are necessary for a subject to match all other colors. In certain abnormal cases it is dichromatic, meaning that only two colors have to be combined to match all colors. This results in an inability on the part of the dichromat to discriminate between certain colors. Among dichromats, there are three known types of this kind of "color blindness": protanopia, deuteranopia, and tritanopia.[4] The first two are characterized by a weakened ability to respond to long-wavelength light (i.e., red-appearing) and medium-wavelength light (i.e., green-appearing), respectively. Both deuteranopes and protanopes have difficulty in discriminating between red-, orange-, yellow-, and green-appearing visual stimuli. Tritanopes are characterized by a weakened ability to discriminate blues or violets from yellows, presumably as a result of a deficiency in their ability to respond to short-wavelength (blue-appearing) light. Let us suppose that we are conducting experiments to determine what type of color vision is present in a hypothetical subject who is known to be dichromatic. We take the frame of discernment to be:

$\Omega = \{$deuteranopia, protanopia, tritanopia$\}$

The power set of Ω is illustrated in Fig. 3.7. Here the bottom-most event represents the empty set, and any connecting line indicates the relationship of being a subset. There are $2^3 = 8$ events in this collection.

The presentation to follow describes how bpa's and belief functions are constructed, and, later combined, as the results from experiments on our hypothetical subject become available. Initially, we have no data other than the hypothesis that our subject is a dichromat, and so we can assign probability only to the certain event Ω. Therefore $m_1(\Omega) = 1.0$, and $m_1(A) = 0$ for all other sets A. Consequently, the belief function Bel_1 gives 1.0 for Ω and 0 for all other sets.

Suppose now that we have conducted an experiment that establishes that in 8 out of 10 trials our subject sees blue and yellow normally but confuses reds, greens, and grays. This makes it reasonable that, to the degree of 0.8, the subject is either a protanope or a deuteranope. Consequently, we can construct a bpa m_2 to reflect these new results by assigning $m_2(\{\text{Deu,Pro}\}) = 0.8$ and $m_2(\{\text{Deu, Pro,Tri}\}) = 0.2$. The corresponding belief function is therefore $\text{Bel}_2(A) = 0.8$ in any set A that contains {Deu,Pro} (with the exception of Ω); the set A could be either {Deu,Pro} itself, or {Deu, Pro, Tri}. Note that Bel_2 is 0 for any set that does not contain *both* Deu and Pro.

Using the combination rule, we can form $m_1 \oplus m_2$; in this case, it is easy to see that $m_1 \oplus m_2 = m_2$. This result makes sense intuitively, because the new bpa m_2 contains the results of a specific experiment, whereas the old bpa m_1 is constructed without any information. Note that there is no way for simple averaging to produce this result, because we cannot obtain $(m_1 + m_2)/2 = m_2$, unless $m_1 = m_2$, which is not the case here.

To resolve whether our subject is a deuteranope or a protanope, we could

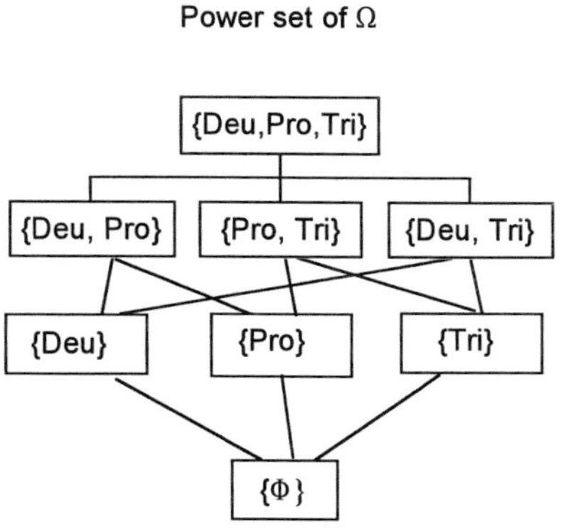

Figure 3.7 The power set of the known types of color deficiencies.

examine the subject's luminosity function. Suppose that by experimental evaluation we determine that the peak of this function occurs at 555 nm. A deuteranope has a peak luminosity typically at 560 nm (which also happens to be the case for normal color vision), whereas a protanope has a peak usually at 540 nm. We could therefore interpret this result to mean that the subject's luminosity curve is three times more likely to be that of a deuteranope than that of a protanope. However, we have to keep in mind that the luminosity test does not distinguish {Deu,Pro} from {Tri}. One reasonable construction of a bpa is therefore as follows: $m_3(\{\text{Deu}\}) = 0.375$, $m_3(\{\text{Pro}\}) = 0.125$, $m_3(\{\text{Deu,Pro,Tri}\}) = 0.5$, and $m_3(X) = 0$ for all other sets X. This construction results in a belief function that gives an equal weight of 0.5 to both the sets {Deu,Pro} and {Tri}, but three times as much weight to {Deu} as to {Pro}.

We can combine the results of the luminosity test with that of the previous experiment by forming the bpa $(m_1 \oplus m_2) \oplus m_3 = m_2 \oplus m_3$ using the Dempster combination rule. Without going through the details, this gives the following result:

$$m_2 \oplus m_3(\{\text{Deu}\}) = 0.375, \qquad m_2 \oplus m_3(\{\text{Pro}\}) = 0.125$$
$$m_2 \oplus m_3(\{\text{Deu,Pro}\}) = 0.4, \qquad m_2 \oplus m_3(\{\text{Deu,Pro,Tri}\}) = 0.1$$

Our overall belief in {Deu,Pro} moves up from 0.8 to 0.9 (the last number is obtained since $\text{Bel}(\{\text{Deu,Pro}\}) = 0.4 + 0.125 + 0.375 = 0.9$). Furthermore, that belief is differentiated between {Deu,Pro} and its elements to favor deuteranopia. These results are intuitively reasonable given the results of the experiments.

An example of the use of Dempster–Shafer theory in computer vision can be found in Safranek et al. (1990). The problem addressed in that work is the updating of a vision-equipped robot's world model from images. A robot may have a model of where objects are located in its environment. To maintain that model in a dynamic environment, the robot can use the following two-step procedure: first, it can predict where the edges in the image would be if the objects are in the positions and orientations in the existing world model; second, it can determine where the edges in the image actually lie by using a standard edge-detection algorithm. The evidence obtained from this second step is used to form a belief function, which is combined with the belief function held by the robot about the validity of the previous position. When the belief in the validity of the world model drops below a certain threshold, the robot must take steps to update the world model.

This concludes our presentation of Dempster–Shafer theory. The next theory to be described, fuzzy logic, originated, like Dempster–Shafer theory, from a need to provide a means for modeling systems about which we do not know enough to construct accurate models.

3.3.3 Fuzzy Logic

The design of a controller for a dynamic system typically requires an accurate model of the system. However, for complex systems there are usually severe limitations to modeling. Zadeh (1973), considered to be the father of the fuzzy logic approach, formalized this idea in what he called the *principle of incompatibility*:

> As the complexity of a system increases, our ability to make precise and yet significant statements about its behavior diminishes until a threshold is reached beyond which precision and significance (or relevance) become almost mutually exclusive characteristics. . . . A corollary principle may be stated succinctly as, "The closer one looks at a real-world problem, the fuzzier becomes its solution." (p. 28)

This principle has been adopted by many engineers to design controllers where imprecision is allowed and, in fact, purposely incorporated to cope with systems for which no precise models exist. One such type of controller, which has applications in data fusion, is the "fuzzy logic" controller. In this section we examine the theory behind fuzzy logic and review its implications for the data fusion problem.

According to the principle of incompatibility, it is fruitless to try to represent the behavior of a complex system solely in terms of precise quantities, such as temperatures or voltages. Rather, it is better to examine to what extent the system behaves in a way that can be described by loosely defined but informative words such as *hot*, *high*, and *near*. Indeed, for many complex systems our understanding has a strong linguistic component but a weak quantitative expression; we can describe what the system is doing in words, but we cannot formulate a set of equations that represent the same behavior.

Zadeh observed that the transition between verbal descriptions and mathematical models is made simpler by allowing for imprecision in the latter. He distinguished between variables that are used *linguistically* and those that are used *quantitatively*: The former occurs when we describe the values of the variable verbally; the latter when we use numbers. For example, *weight* is used linguistically when its values are described as "heavy," "not too heavy," "light," or "very light," and quantitatively if we take its values to be measured in kilograms. Linguistic variables can be represented by functions that are defined on the domain of allowable values of the variable. For example, if weights between 0 and 100 kg are possible, we can describe the extent to which an object x is heavy by a function μ that takes values between 0 and 1 for x between 0 and 100, the value $\mu(x)$ indicating to what degree the object is heavy. Then $\mu(x)$, which is called a *membership function*, may take the form indicated in Fig. 3.8.

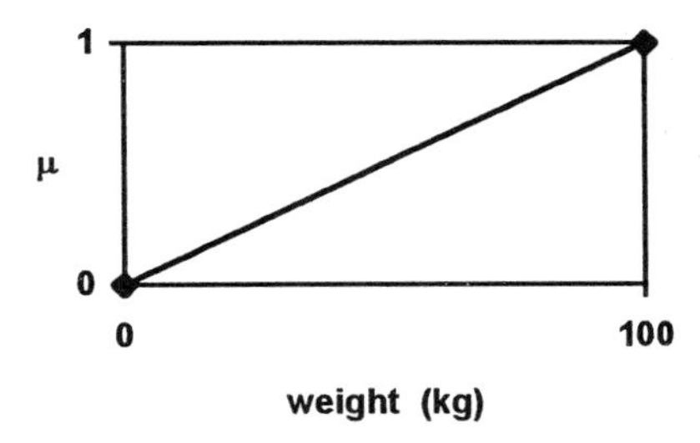

Figure 3.8 The membership function for the property "heavy."

This function gives a linear association between the weight of the object and its heaviness. We could also define suitable membership functions to represent the degree to which an object is "light" or "medium" in weight. Typical functions are shown in Figs. 3.9 and 3.10.

A logic can be developed that works on membership functions rather than on the variables themselves. Here the term *logic* means a system of reasoning that includes implication, Boolean operations like union (OR) and intersection (AND), and modifiers like "very", and "nearly." This logic of membership functions is what has become known as fuzzy logic. We now examine the elements of this logic. (Readers who would like a more detailed explanation are referred to Mendel, 1995.)

Membership functions can be modified by the use of operators. For example, the modifier "very" makes values of the membership function that are near 1.0 more significant. This can be modeled by the operator $\mu(x) \mapsto \mu(x)^2$. This operator when applied to the membership function for "heavy" produces the result for "very heavy" shown in Fig. 3.11. Similar modification operators can also be defined for "not," "extremely," "more or less," etc.

Combination operators between membership functions are of the most interest to us because of their role in data fusion. Fuzzy logic allows the full range

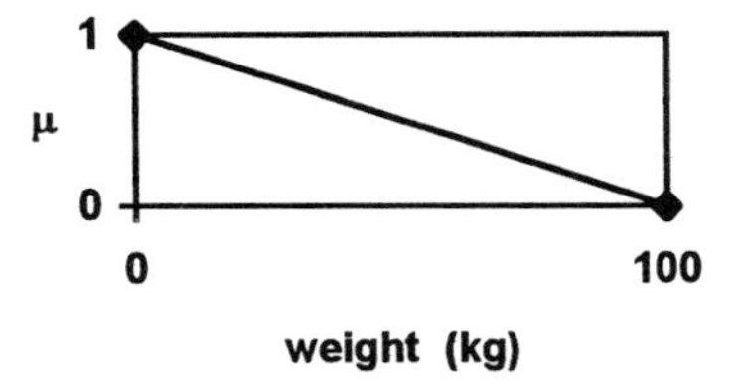

Figure 3.9 The membership function for the property "light."

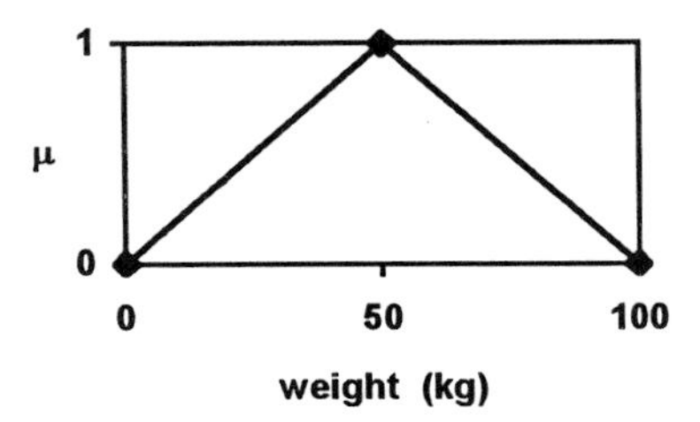

Figure 3.10 The membership function for the property "medium."

of Boolean combination operators. The intersection (or logical AND) between two membership functions is defined as follows:

$$\mu_1 \cap \mu_2(x) = \max[\mu_1(x), \mu_2(x)]$$

Similarly, the union (logical OR) is defined as:

$$\mu_1 \cup \mu_2(x) = \min[\mu_1(x), \mu_2(x)]$$

To illustrate these rules, let us examine the membership function for "either light or very heavy": using the graph for "very heavy" shown in Fig. 3.11, the result is shown in Fig. 3.12.

Fuzzy logic has already been applied to the problem of image segmentation. Huntsberger et al. (1985) describe an iterative segmentation scheme that incorporates information from red, green, and blue components of an image. On its own, each component is insufficient for accurate segmentation, and the simple "gray-level" combination of all three components does not produce good results either. A data fusion approach would attempt to combine the segmentation obtained from each component. The method proposed by Huntsberger et al. to do just that is to partition the space of (R, G, B) values into c regions. Each region has a corresponding membership function, so every pixel has c membership

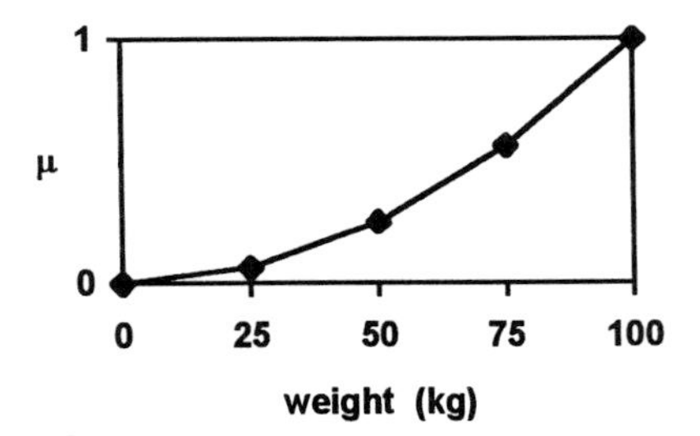

Figure 3.11 The membership function for the property "very heavy."

function values. The partition is iteratively refined so that the pixels have the maximum sum of membership values; therefore, their degree of belonging to a single region is maximized. The membership values are calculated in terms of the geometric distance of the (R, G, B) values of a pixel to each of the c centers of the regions. The iteration produces a "fuzzy clustering" of the colors around c locations in color space. A threshold applied to the membership values is then sufficient to produce the overall segmentation.

This concludes our discussion of fuzzy logic. The next section presents a most recent theoretical development: Lebesgue logic.

3.3.4 Lebesgue Logic

The logic of probability measures called *Lebesgue logic* was developed by Bennett et al. (1993). Its origins lie in a formal theory of perception, called *observer theory*, that was proposed earlier by Bennett et al. (1989). Observer theory is briefly reviewed here first because its development motivated the study of the new Lebesgue logic for probability measures. As is discussed later in this chapter, this new logic is a promising approach for data combination in vision and otherwise.

The central thesis of observer theory is that all perceptual capacities, e.g., stereo vision, have the same underlying structure when viewed in information-processing terms. This structure is called an *observer*, and it is illustrated conceptually in Fig. 3.13. The symbols used in this figure have the following meanings: Y is the set of all possible premises for the perceptual capacity being modeled; e.g., for stereo vision, Y would be the set of all possible left and right views of a scene. X is the set of possible conclusions; for stereo, this would be the set of all possible left and right views in Y, with the corresponding depth maps for each pair of views. The process of obtaining premises is via a mapping π that takes conclusions in X and maps them onto Y; in stereo vision, this mapping is the removal of depth information to produce only two-dimensional

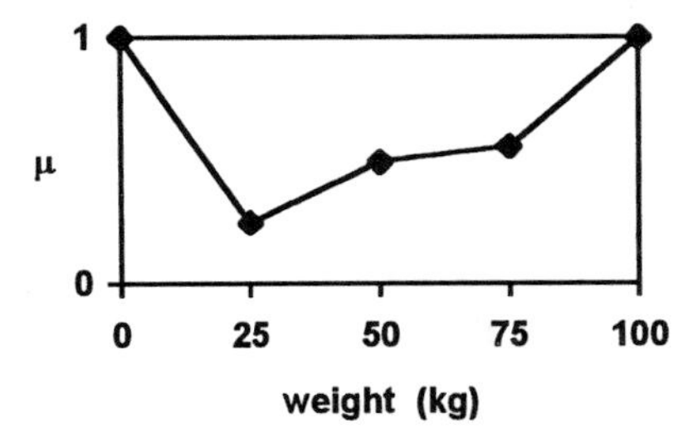

Figure 3.12 The membership function for the property "either light or very heavy."

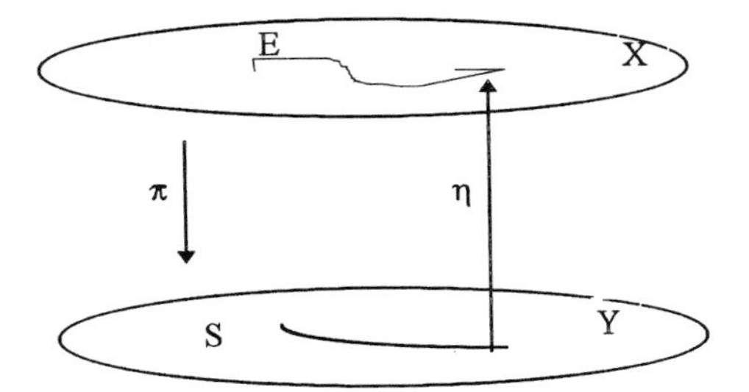

Figure 3.13 A conceptual diagram of the hypothetical observer. (After Bennett et al., 1989.)

left and right views. The perceptual system's job is to recover a depth map from a premise in *Y*. It may not be possible to recover depth for all pairs of images. Therefore, the symbol *S* is used to represent the subset of *Y* of image pairs where depth recovery is, in theory, possible; the symbol *E*, which is "upstairs" in *X*, represents the set of recoverable image pairs plus their depth maps. The process of going from premises in *S* to conclusions in *E* is represented by η, which for each premise in *S* provides a probability measure on *E*. Perceptual conclusions are modeled in observer theory as probability measures, because for a single premise it may be possible to reach more than one conclusion. For example, we can reach more than one three-dimensional (3-D) interpretation for the two-dimensional (2-D) Necker cube illustration.

Detailed examples of observers representing the perceptual capacities of structure-from-motion, stereo vision, and the visual detection of light sources are given in Bennett et al. (1989). Here, we focus on the representation of premises and the Lebesgue logic (Bennett et al., 1993) that emerged from the observer mechanics approach.

Each conclusion given by an observer is a probability measure on the observer's space of possible conclusions *E*. The conclusion of one observer could serve as the premise for another, higher-level observer. For example, suppose that we were to model the combination of depth information obtained from structure-from-motion (SFM) with that from stereo vision, both of which are capacities that our visual system possesses. The SFM observer would give its conclusion as a probability measure on the space of possible 3-D surfaces. Similarly, the stereo vision observer would also give a probability measure on 3-D surfaces, but one that is likely to be different from the measure given by the SFM observer. It is possible to imagine a higher-level "3-D surface" observer that obtains on its space of premises both the 3-D surface probability measure from SFM and the one from stereo. How are these two measures to be interpreted? When does one measure "imply" another? And what might be their logical

AND or OR? These questions motivated the development of Lebesgue logic for probability measures, which is now described.

Let us suppose that we have two probability measures, denoted λ and ψ, on Y. These two measures represent information obtained from two different sources about the same perceptual event. Then one measure, say, λ, is deemed to imply the other, ψ, if there exists an event A such that

$$\psi(B) = \frac{\lambda(A \cap B)}{\lambda(A)}$$

for all events B. Intuitively, this means that ψ is the restriction of λ to the event A. The significance of the imply relationship is that if λ implies ψ, then there is essentially no new information in ψ to that already provided by λ. The two probability measures are deemed to be *simultaneously verifiable* if there exists a nontrivial event A on which they are the same measure, up to multiplication by a scalar. More precisely, there must exist a constant K such that for every event B:

$$\lambda(A \cap B) = K\psi(A \cap B)$$

This equation can be understood intuitively to mean that if the universe is restricted to A, then λ and ψ are the same measure, up to a constant factor. For two simultaneously verifiable measures, the AND is defined to be the measure obtained by restricting either λ or ψ (it turns out not to matter which) to the set A and normalizing:

$$\lambda \wedge \psi(B) = \frac{\lambda(B \cap A)}{\lambda(A)} = \frac{\psi(B \cap A)}{\psi(A)}$$

When defined in this way, $\lambda \wedge \psi$ is the "largest" measure that is implied simultaneously by both λ and ψ; i.e., if there is any other measure δ that is implied by both λ and ψ, then $\lambda \wedge \psi$ implies δ. Similarly, the logical OR of λ and ψ, denoted $\lambda \vee \psi$, is defined to be the "smallest" measure that implies both λ and ψ; i.e., if δ implies both λ and ψ, then δ implies $\lambda \vee \psi$ as well.

Example 1. This example is adapted from Bennett et al. (1993). Suppose that Y consists of just four points, denoted a, b, c, and d. Let λ and ψ be the two measures shown in Fig. 3.14. It is clear that these measures are simultaneously verifiable, for they agree on the set $A = \{b,c\}$. Consequently, we may form the AND, denoted $\lambda \wedge \psi$ as earlier, by restricting either λ or ψ to A (it does not matter which) and normalizing the result to make it a true probability measure. The result is shown in Fig. 3.15. Also shown, for comparison, is the OR, denoted $\lambda \vee \psi$, which is formed by constructing the smallest measure that is implied by both λ and ψ.

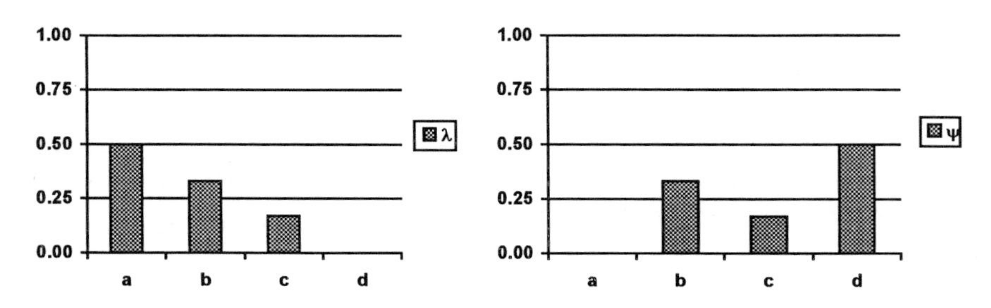

Figure 3.14 The two probability measures for the functions λ and ψ.

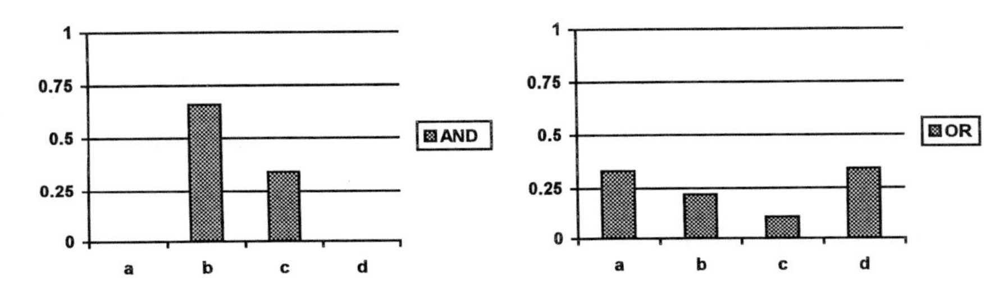

Figure 3.15 Normalization of the functions shown in Fig. 3.14.

Example 2. This example is also adapted from Bennett et al. (1993). From stereo views of four noncoplanar points in space, one can in theory infer the depths of all points up to an unknown scale factor that depends on the fixation distance. Therefore, a stereo observer would be able to return as its conclusion a probability measure on the one-parameter family of interpretations corresponding to the four points. Now, if these same four points are observed in rigid motion, then from three distinct views it is theoretically possible to determine the depth values up to a single mirror reflection. Therefore, an SFM observer would return as its conclusion a probability measure supported only on the two possible solutions. The two SFM solutions do not both fall into the one-parameter family of solutions obtained from stereo vision; as Richards (1984) has shown, only one of the SFM solutions does. Consequently, it is possible to obtain a unique solution for the 3-D structure of these four points by using *both* stereo and SFM, but it is not possible to do so from each individually. This, of course, is the fundamental concept driving this entire book.

In terms of Lebesgue logic, this combination of probability measures is interpreted as the logical AND of two probability measures. Each measure is defined on the space of possible 3-D structures with the four points. The set A in this case (see the equation on page 93) consists of the single 3-D structure that lies in both the SFM solution set and the stereo solution set. The AND of the two measures is simply a probability measure that gives unity weight to the structure and zero weight to all others.

Observer theory provides a means for implementing the logical combination of measures from the Lebesgue logic. Suppose we have a single probability measure λ on an observer's Y space, which may be the result of a combination of two or more probability measures from lower-level observers. The measure λ now represents the observer's subjective assignment of probability to the possible premises. If the observer has no uncertainty about the premise, then λ would be concentrated at exactly one point $\{y\}$ in the space Y. The structure of the observer now allows us to calculate what the interpretations of the observer are for each possible premise λ. Recall that the interpretations are modeled by the kernel η, which for each element of S gives a probability measure on the set of corresponding interpretations in E. The probability of interpretations, denoted here μ, is then obtained as the superposition of all of the probability measures contained in η, each measure being weighted by the corresponding probability on premises given by λ. For example, suppose that S contains only two elements, s_1 and s_2, and that the probability measure given by η for each is denoted $\eta(s_1,\cdot)$ and $\eta(s_2,\cdot)$. Then the probability measure μ is simply

$$\mu(A) = \frac{\lambda(s_1)\eta(s_1,A) + \lambda(s_2)\eta(s_2,A)}{\lambda(s_1) + \lambda(s_2)}$$

The denominator is a normalizing term, which is used to make sure that μ is a proper probability measure so that $\mu(Y) = 1$.

This concludes the discussion of Lebesgue logic. The next section compares the approaches and methods of the theories that have been discussed in this section.

3.3.5 A Comparison

The formal mathematical theories presented in this section each provide a language for formulating and analyzing data fusion problems. The concept of a probability measure is fundamental to two of them: Bayesian statistics and Lebesgue logic. The power and elegance of probability theory can be brought to bear on data fusion problems formulated in either of these two theories. However, a problem emerging whenever one uses probability measures is that ignorance must be modeled by a uniform measure. This constraint can lead to nonsensical predictions, as we have shown earlier. Both Dempster–Shafer theory and fuzzy logic eschew probabilities in their models for this reason. However, a basic probability assignment as used in Dempster–Shafer theory is still similar in some ways to a probability measure. For example, it must sum up to 1 when added over the power set. In contrast, a membership function such as that used in fuzzy logic does not need to satisfy any constraint other than being positive-valued, with a maximum value of 1.

It is now appropriate to compare the combination rules for Lebesgue logic, fuzzy logic, Dempster–Shafer theory, and Bayesian statistics. Lebesgue logic and Dempster–Shafer theory both require that two propositions be compatible before they are combined; in the former they have to be simultaneously verifiable probability measures, and in the latter they have to be bpa's that do not assign all their weights to disjoint events. These compatibility relations are similar, for they both rule out combinations of inconsistent information.

On the contrary, the odds-updating rule of Bayesian statistics and the logical-combination rules of fuzzy logic allow arbitrary combination. Bayesian updating ignores the consistency of data. In fuzzy logic, the logical AND and OR can also be applied to any two membership functions, no matter how inconsistent.

Fuzzy logic, Dempster–Shafer theory, and Bayesian statistics have been applied to data fusion in image processing and computer vision. To date, there has not been an application of Lebesgue logic, although the methods provided there suggest a powerful and novel approach and may provide an interesting alternative to some of the others already applied to this important problem in computer and organic vision.

3.4 CONCLUSION

This chapter has reviewed a few of the many models of vision in which the combination of separate cues, attributes, or elements of the stimulus scene plays a central and important role. We believe that combination, or data fusion, is a fundamentally important concept in the study of vision, whether it be a purely artificial computational problem to be solved or a search for an understanding of how the organic system has evolved that motivates the study. Indeed, the evolution of the organic system may have been driven by the same forces that are emerging in the computer science field as necessary and efficient for achieving effective image understanding.

Clearly this is a major shift in thinking, in both biopsychology and computer science. In the biological and psychological sciences, it has long been traditional to deal with a single variable in isolation from others that may influence the same visual experience. It seems obvious now, with many studies showing that there are powerful interactions between stimulus attributes, that this approach is, at least, obsolescent and has, at worst, terribly misled us in our general understanding of the way the organic visual system works. As we saw in Chapter 2, combination and interaction are biopsychological realities.

It is even interesting to appreciate that many of the older criteria for evaluating theories may no longer be meaningful in the context of organic vision. Simplicity, parsimony, and economy need no longer guide and constrain our theoretical explanations. In a system with an abundance of components, redundancy and dependability may be the criteria of choice.

The history of biology and psychology is reflected in the current status of the field of computer vision. Heretofore, most development efforts have been directed toward the refinement of algorithms and operators that manipulated only a single attribute of the acquired image. We believe that there are fundamental limits to such refinement, and the only hope for the future lies in the realm of combination methods of the kind discussed in this chapter.

Having now presented this review of some of the more notable past, present, and even future theoretical approaches to combining images, we turn to some of the specific developments that have come from our laboratory. The remainder of this book presents a discussion of the current status of the SWIMMER project—our vision system. The next chapter presents an overview of our computational model, updating and adding new developments to the discussions presented in our earlier work (Uttal et al., 1992). Chapters 5 and 6 present the specific details of our combination methods for two- and three-dimensional images, respectively. Chapters 7 and 8 deal successively with the pattern recognition and surface reconstruction techniques that are also essential parts of our project.

NOTES

1. R. Kakarala would like to thank Gary Prentice for helpful discussions during the writing of his sections of this chapter.
2. Some of the material in the following section describing Marr's model of three-dimensional perception is adapted from Uttal (1988).
3. This is a consequence of Glivenko's theorem (see Bronshtein & Semendyayev, 1973, p. 621, where it is given the grand title "main theorem of mathematical statistics") that the empirical distribution of a sample tends, as the size of the sample increases, to the true distribution of the underlying random variable.
4. The Optical Society of America's Committee on Colorimetry volume entitled *The Science of Color* (1953, p. 137) identifies a fourth type of dichromatism, called *tetartanopia*, but it is apparently exceedingly rare.

4

A Vision System

4.1 INTRODUCTION

So far in this book, we have presented background philosophical, neurobiological, and psychophysical material to set the stage for the technical computer science discussions that follow in this and later chapters. It is the purpose of this chapter to describe the general organization of our vision system—the computational model we call the SWIMMER. The simulated SWIMMER "exists" in a complex three-dimensional microworld that requires a variety of programming activities to instantiate. The SWIMMER is designed to be capable of acquiring images of food objects, recognizing and discriminating among regular (edible) and irregular (inedible) objects, establishing a three-dimensional world model of its environment and the objects in it (including itself), and then demonstrating its understanding by swimming through a turbulent ocean to those objects. Involved in such a simulation are program modules that simulate the visual, localization, interpretative, decision-making, and motor functions of the SWIMMER.

The material presented in this chapter consists of several parts. First, the general organization of the model is discussed, to indicate how the components described in later chapters fit together conceptually. This section also discusses our general approach and how the computer model corresponds to the neurobiological, psychophysical, and computer science discussions presented in Chapters 2 and 3. Second, we describe how new developments in automatic programming have expedited our work and replaced the more primitive system we developed and described in the previous book (Uttal et al., 1992). Third, this chapter describes some supplemental developments of the visualization aspects of our system above and beyond those reported earlier.

The vision system we call the SWIMMER has developed far beyond some of the primitive notions that were described in the first book in many of its aspects. Some of these developments have occurred because of the general improvement in the science of computational modeling, and some have been contributions based on the efforts of all of us who worked on this project. Clearly, there is a symbiotic relationship between our work and that of other laboratories. Nevertheless, each laboratory has its own particular style and goals. Another role of this chapter is to clarify what are our particular goals.

We have been developing a theory of vision based on combination and fusion. Everything that we have done, whether directly or indirectly connected to this goal, is aimed at instantiating the fundamental logic of premises I and II—the segregation of the information representing scene attributes into separate channels and the recombination of that segregated information at some higher level of the brain or later step in the computer program, respectively. This goal is the raison d'être of our project and the foundation rationale that has guided all of our work.

However, it was also deemed of value to concretize our work by embedding it a practical application. For us, that application was the SWIMMER, the simulation of an autonomous, visually guided, underwater predator. The SWIMMER system took many forms. These included: (1) its manifestation in our plans, expectations, and understanding of how vision worked; (2) a large computer programming system; and (3) even a real connection between a small, remotely operated vehicle (ROV) and the computer control programs. Unfortunately, this last embodiment was never completely fulfilled. Although we able to bring the ROV under the control of a computer program and to process video information from the ROV, we were not able, before the ROV became unavailable to us (for administrative reasons that are not germane to the present discussion), to connect the control signals generated by the vision system to the ROV. In place of that real physical embodiment, we went on to simulate the behavior of the entire system in an elaborate computer program and visualized the operation by means of dynamic computer displays.

There is another aspect of our work that we wish to acknowledge to a degree that is not, we believe, usual in this field. That is, the full-blown version of our system, with all of the bells and whistles, does not operate in real time. In the biological domain of the organic visual system, the parallelicity of action provided by the huge number of available neurons brings enormous neural computational power to play and, thus, virtually real-time responses to even complex visual stimuli. Immediate interpretation of complex scenes and adaptive response selection can occur even though neurons may have speeds many orders of magnitude slower than computer logical units. The reason for this is obvious. The brain has 10^{13} neurons. The typical laboratory computer has but one. In the standard computer available to us (SGI Indigo^2 units with 150-megahertz operating speeds) this single arithmetic unit required that all instructions be executed in serial order. Thus, even though the fundamental speed of the computer was 150,000 times faster than a typical neuron, the huge number of instructions that had to be carried out meant that it often took frustratingly long to complete a simulated experiment.

It should not be overlooked in this context that we were often surprised when we applied some algorithm published in the literature (usually without the author's having mentioned this caveat) to find that it, too, required inordinate amounts of time for its execution. Speed of execution, however, is merely a practical problem, one that will ultimately be solved by faster or more parallel computer systems or ingenious, fast algorithms; future engineering developments will resolve this impediment to the application of vision systems such as ours to real-world problems.

Nevertheless, there is a related problem that is fundamental and that cannot be deferred to the future: The attribute segregation and combination process is and must continue to be inherently slower than any procedure depending on only a single attribute. This is not a handicap that can be overcome without the extreme parallelization of functional centers (as opposed to the parallel organization of neurons within a center), comparable to those found in organs like the brain. It is, however, unlikely that either necessary kind of computer parallelization will be available for many years. Thus, our model is presented as the embodiment of an unusually comprehensive conceptualization of a vision system rather than as a working tool.

Indeed, any artificial-intelligence project of the kind we describe here would be presumptuous if it assumed that the level of competence of such a simulated computational system comes close to imitating the real biological system. At best, any simulated vision system is a poor approximation to some of the early stages of image processing carried out by the visual system of even a simple organism. The best computational model of this kind (and we believe that ours comes closer than any other yet reported to achieving such a performance level) incorporates only a small portion of the cognitive capabilities of the human visual system. Given the enormous and still-underappreciated impact of *cognitive penetration* (i.e., the influence of additional top-down, information-based knowledge not contained in the original image) into human visual perception, our system can be considered only a crude model of the early, preattentive, and more or less automatic stages of information processing by organic nervous systems. It is only when we simulate the pattern recognition and reconstructive processes that our work begins to approximate some of the higher-order aspects of human vision, and, even then, in ways that most likely incorporate logical functions quite different from those used by the brain.

Like all other models of vision, the range of visual images that can be processed is constrained to a limited universe of discourse—only certain types of images can be processed. To extend that universe requires that additions and modifications be made to the programs; they typically do not generalize far beyond the type of image on which they were tested. Our model, therefore, must be understood to be an attempt to simulate the global organization and the macrofunction of a vision system rather than the microdetails of the logic underlying those functions.

In spite of these limitations (which are common to all computational models or theories of vision), we believe that we have made some progress toward understanding the principles by which vision operates. To reiterate, our central thesis is that it is *only* by combining information from the separate attributes of the visual image that even a remote approximation to full-fledged vision is possible. This is the core idea of this book; our computational model of a complete vision system is one way of making this important point.

In the remainder of this chapter, we provide a general introduction to the SWIMMER model, discuss some of the nuts-and-bolts details of implementing such a system, and consider the visualization package that has allowed us to observe the process of the model, both in terms of its computational steps and in terms of the generated behavior of the SWIMMER.

4.2 THE GLOBAL STRUCTURE OF THE SWIMMER

The SWIMMER is a much more complete vision system than is typically presented in the literature. One has only to peruse the computer science journals to appreciate that integrated (i.e., multiattribute) theories are rare. Our model, on the other hand, considers most of the obvious attributes of a visual scene. It separates and then combines intensity, color, texture, disparity, motion, shading and contour attribute information.

Figure 4.1 is a block diagram of the entire system in its current version. (This block diagram should be compared with Fig. 1.1 from the original SWIMMER book, Uttal et al., 1992, to appreciate the progress that has been made in the interim.) Such a comparison would show that most of our recent effort has been dedicated to the improvement of the combination, recognition, reconstruction, and visualization processes. The remaining chapters of this book document the specific changes made in these parts of the vision system. However, this section of this chapter emphasizes some of the more global strategic changes that have been implemented as the model evolved.

One of the most important additions to our model has been the addition of

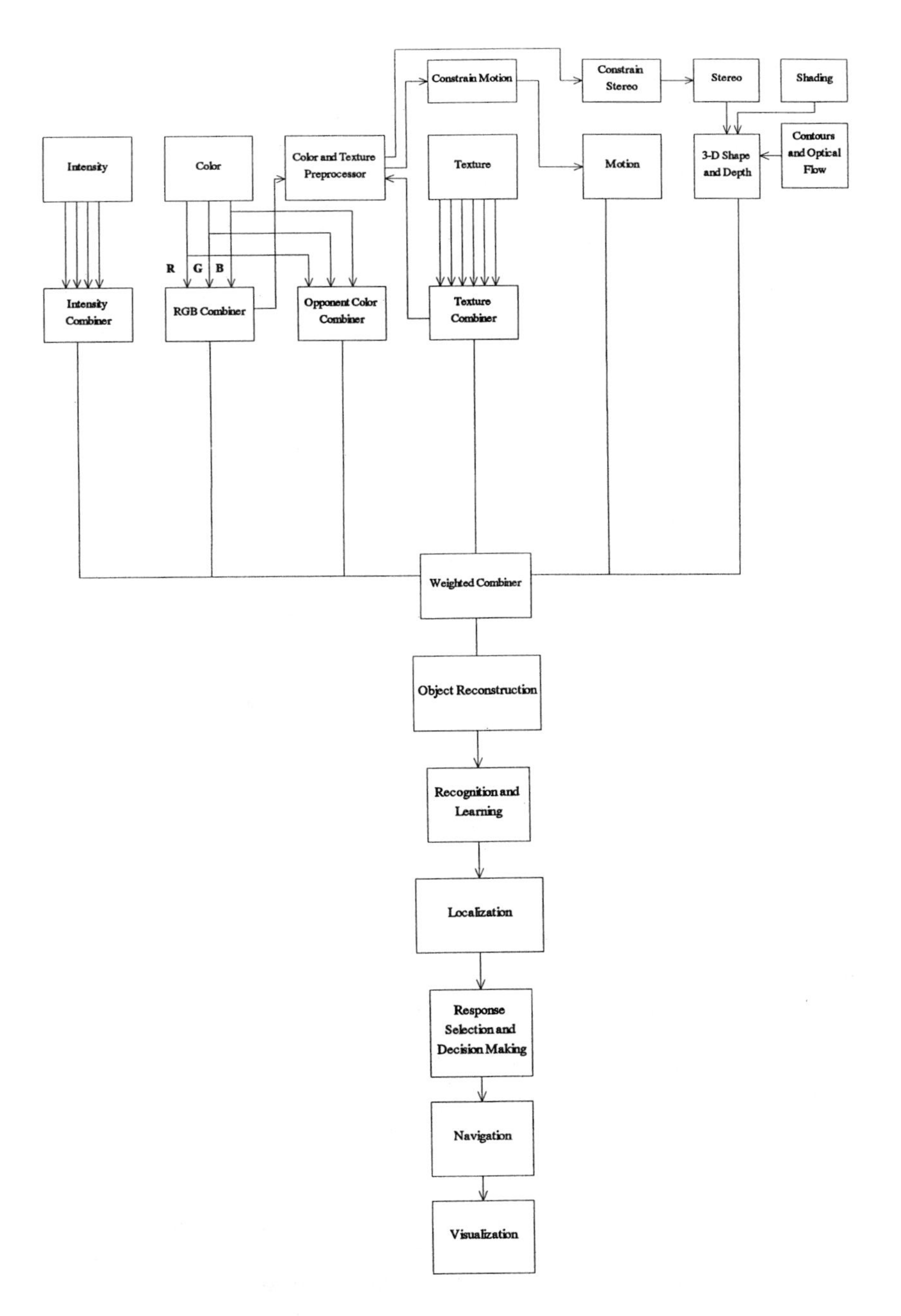

Figure 4.1 A block diagram of the entire computer vision system that was developed in our laboratory.

lateral constraints. This is also, of course, another kind of combination process. In our original model each attribute of the initial image scene was processed independently until a final, weighted-combination stage brought the results of processing each attribute together. For example, although there might be several algorithms combined to determine texture boundaries, texture operators were initially combined only with texture operators, color with color operators, and so on. The combination process, in other words, was carried out only within the confines of each attribute until the final, weighted combiner.

Isolating the combination processes to only a single attribute until the last stage of processing is not necessarily an efficient process, nor is it true to what we know about the organic visual system from both psychophysical and neurophysiological studies. The rule in the organic world is interaction and combination across and between as well as within the various attributes. It is this interaction that helps to resolve ambiguities and allows organisms to solve visual problems that may initially seem to be unsolvable. Therefore, in the updated version of our vision system we have added methods for *interattribute* combination to the *intraattribute* combining techniques installed in earlier versions of our model.

There were three ways in which interattribute combination was implemented. In the first, two-dimensional boundaries determined by the color, texture, and intensity attributes of the scene were combined to give a better two-dimensional solution to the segmentation task. This was essentially where we stood when the previous report (Uttal et al., 1992) on our project was published. In that earlier work we had concentrated on what we called *spatial averaging* to provide the combining mechanism. Spatial averaging involved a simple statistical averaging of the positions of the two-dimensional boundaries obtained from a set of six edge detectors, specifically those sensitive to texture. Chapter 5 in the present book offers a novel alternative approach to two-dimensional combining in which a model borrowed from particle physics is applied as the combining mechanism. The particle model uses boundary information from all three attributes to solve the two-dimensional segmentation problem posed by an image.

Second, robust two-dimensional information was combined with three-dimensional information to give a better estimate of the surface shape or depth of an object. In particular, color and texture were combined to provide a constraint to stereo and motion algorithms.

The third way in which the combination philosophy was applied, however, was the most ambitious. We extended this key and central idea of combination to solid shapes: Two or three estimates of the three-dimensional shape of a surface were combined to produce improved and more robust estimates of surface shapes. Chapter 6 presents a detailed discussion of how this task was accomplished.

In addition, our work on two-dimensional recognition progressed in a satisfactory manner. Chapter 8 describes important steps that have been added to the size-, orientation-, and position-invariant shape recognizer previously described in Shepherd et al. (1992). This new work is especially interesting, because it extends the recognition component of the model so that it is capable of recognizing incomplete or occluded objects. Finally, Chapter 9 describes how incomplete surface shapes can be recovered by vision systems by means of some novel interpolation algorithms in a way that models human perceptual filling-in processes.

The important aspect of all of our work, however, is that is integrated into a unified computational system. The system is conceptually complete, even though all of the components have not yet been developed to a level that satisfies us. Furthermore, it is modular. That is, the system has been developed so that the individual blocks shown in the block diagram in Fig. 4.1 can be removed and replaced with improved or up-to-date versions as they become available. This modularity was expedited in recent years by the availability of the IRIS Explorer programming system described in the next section.

4.3 THE APPROACH

We have already spoken of the significant initial philosophical commitment made to a particular programming approach at the beginning of this project—initially to segregate attribute information and then to recombine it at a later stage of the process. However, there were also several other important choice points at which decisions were made that significantly affected the course of the project in major ways. Some have become so deeply embedded in our work that it was easy to ignore them on a day-to-day basis; yet some of these decisions had tremendous consequences for our activities in terms of the long range development of our vision system.

One of these important initial decisions led to a commitment to eschew neural-network programming. Although a plausible system could have been constructed using such an approach, it was our conclusion after extensive consideration of this issue that many of the desired general goals as well as the modules to be developed were beyond the capabilities of a neural-network approach. This was simply a practical decision based on our estimate of the relative power of this technique compared to ordinary analysis or computational modeling. There was, however, a deeper reason that led us not to use neural-net algorithms. To put it baldly, although we accepted the fact that computational neural-net methods and biological neural nets exhibit some superficial similarities (such as distributed nodes, parallelicity, and changes in synaptic efficiency and weighting as a function of experience), we did not believe that the current neural-net models are any more valid theories of the logic or inner workings of the biological system than are ordinary mathematical formulae or conventional computational algorithms.

There are two reasons for this conclusion. The first is that actual biological networks and neuronal logic are still opaque to the kind of structural analysis that is implied—but not realized—by the neural-net approach. The second reason is that the neural-net technology and programming art, as it currently stands, is not adequately linked to the organic neural nets. What has happened is that the general idea of parallelicity has been applied as a guide to some ingenious computer programming to simulate or analogize certain behavioral functions. However, we see this as more of an engineering tour de force than a valid biological theory. Readers interested in a more detailed discussion of this argument are referred to Uttal (1998).

Another major decision that has had enormous implications in the course of the development of our vision system occurred when we decided to push ahead with the development of a modular model, in other words, one in which relatively isolated and discrete algorithmic components could be developed independent of each other. Then, with minimum difficulty, these independent modules could be interconnected in novel arrangements or even completely replaced when improvements were made. Originally, the ability to modularize and substitute modules was embodied in what we called the "Menu-Driven Executive System," as described in the first book on the SWIMMER project (Uttal et al., 1992) Modern developments in programming systems design by commercial suppliers, however, quickly surpassed our efforts to develop such a utility system. In particular, The IRIS Explorer Graphical Programming System was offered by the Silicon Graphics Corporation for use on our system. The Explorer system proved to be a powerful tool implementing exactly the features we needed and replaced the "Menu Driven Executive System" as soon as it could be loaded into our operating system. The following section briefly describes the IRIS Explorer system and discusses its implications for our project.

4.3.1 The IRIS Explorer System

The IRIS Explorer system provided on SGI computers is a modular system for developing and visualizing the output of complex systems of computational modules. The system allows both experienced programmers and nonprogrammers to assemble complete programming systems from a library of previously prepared modules. The selection and assembly process is controlled very simply by manipulating the computer mouse. Each module is "dragged" from a *Module Library* to another computer window, designated as the *Map Editor*, simply by selecting it and then moving the cursor to the desired position. The Map Editor visually depicts the current organization of the system under development and permits control of parameters, also by means of the mouse.

Systems of programs may be assembled in this way without any detailed programming on the part of the programmer—*if* the functions to be carried out had previously been incorporated into the Module Library as IRIS-Explorer-compatible library modules. There are two sources of library modules: (1) The IRIS Explorer system came with a large library of its own preprogrammed modules. (2) It was possible for us to develop our own modules according to a relatively standard protocol. In our system only a few of the IRIS Explorer library modules were suitable, and most of the modules that made up our vision system were programmed and then adapted into the library by us.

When an Explorer-style module is constructed, it must follow strict restrictions on its input and output conditions. This forced standardization of module construction and organization had three important and advantageous effects on our work. First, the standardization made it possible for most modules to be used in many different versions and many different parts of the vision system. Second, standardization expedited the replacement of modules in our model whenever they were supplanted by more efficient or effective versions later in the course of the project. Third, standardized input and output requirements also permitted the quick assembly of special-purpose versions of the system to test quickly some new idea or to satisfy the needs of some special application.

Once placed in the Map Editor by the programmer, the selected modules are interconnected by dragging virtual lines (again with the mouse) between the selected modules. These lines acted as channels controlling the flow of information among and between the modules. To establish such a channel, the programmer had only to tag the output region of one module with the mouse and then to move the cursor to the input region of the module representing the next step in the program. This process cannot be carried out with complete abandon—the input to each module requires a compatible output from the preceding module. The exceptions to this generalization, of course, are modules that generate an original image (and thus that have no input) or those that display or render the output and thus are followed by nothing other than the computer window display system. (As it turned out, most of the difficulty in using the IRIS Explorer system was encountered in matching module outputs to subsequent module inputs when we created new modules.)

Figure 4.2 is an IRIS Explorer map of a portion of the vision system shown in the block diagram of Fig. 4.1. It illustrates some of the attributes of the IRIS Explorer module and also represents an alternative representation of at least this part of the vision system. This map is made up entirely of modules that were developed in our laboratory. The *logoVII* module acquires a stored image that had previously been captured by the video processing equipment attached to the computer and stored in the computer's memory. The video processing equipment consisted of a pair of CCD cameras connected to the computer through an SGI Galileo video controller and circuit board. Images were, thus, acquired and

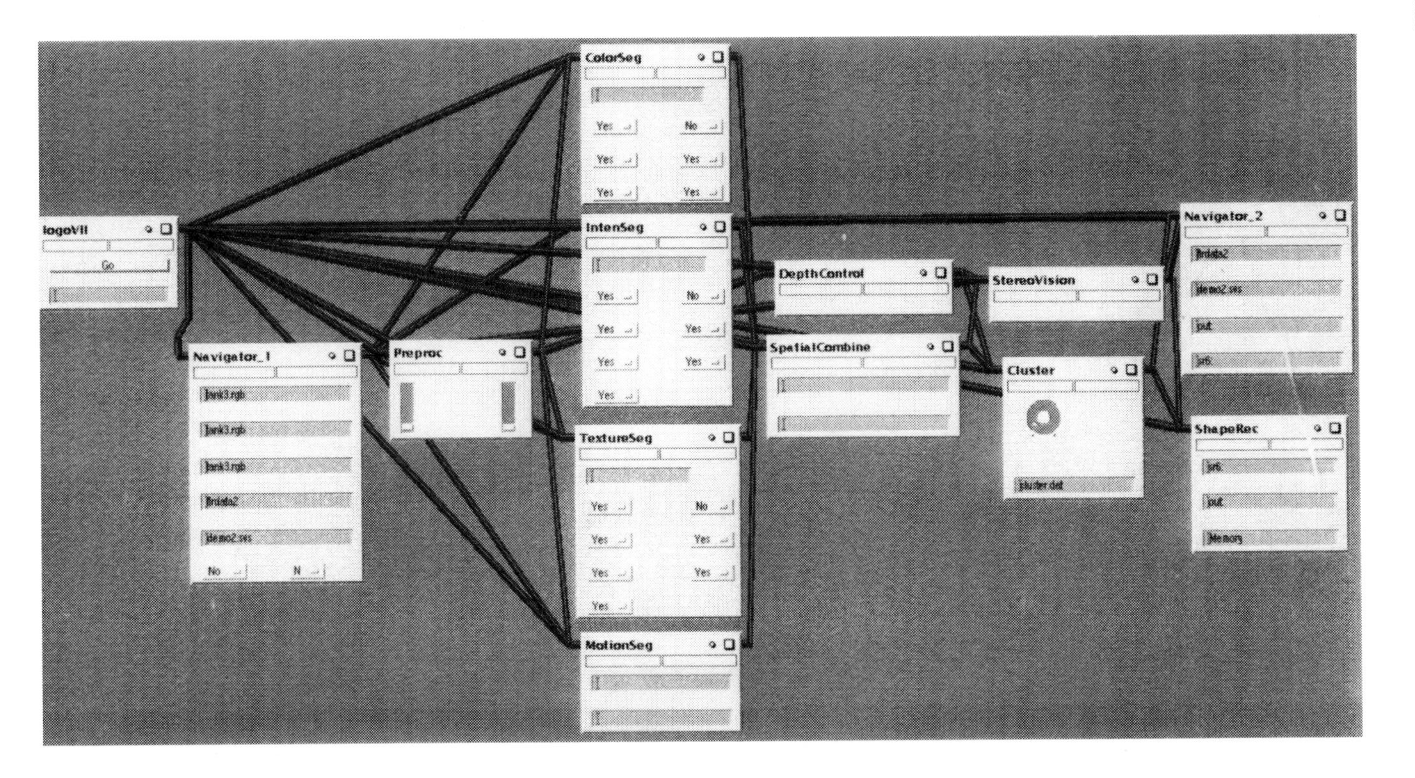

Figure 4.2 An Iris Explorer map of the central portion of the vision system shown in Fig. 4.1.

stored in the computer memory as inputs to the Map Editor representing the sequences of processes incorporated into a given version of the SWIMMER model.

The *Navigator 1* module controls the initial phases of the visualization of the simulated SWIMMER as it goes about its operations. This module initially displays search behavior on the computer screen as if the SWIMMER was looking about to discover any objects in its simulated world. *Navigator 1*, therefore, is the locus of the world model in which the environment of the SWIMMER is encoded. The *Preproc* module cleans up and prepares the input image for the four segmentation modules based on the color (*ColorSeg*), intensity (*IntenSeg*), texture (*TextureSeg*), and motion (*MotionSeg*), respectively. Each of these segmenters operates on the same single multiattribute input image, but each in its own way according to its own sensitivities.

The heart of this IRIS Explorer map and the philosophical core of our vision system, of course, is the combining module (*SpatialCombine*), which collects the information from all four segmenters to produce a robust estimate of boundary information. In some cases, the output of *SpatialCombine* produces fragmented or broken outputs. The *Cluster* module was designed to clean up incomplete images by attaching fragments to each other that are within certain threshold distances and that otherwise shared common properties.

The stereo vision module (*StereoVision*) operates on a pair of pictures provided by *logoVII* to solve the correspondence problem and then to compute the depth of the images. (See Chapter 6 for the details of this part of the vision system.) The *DepthControl* module is one type of constraint applied to improve the performance of the *StereoVision* module. (Not all of the modules shown in Fig. 4.2 were present in every version of the vision system. For example, *Preproc* and *Cluster* were not present in every version; in some other situations it was even the case that not all of the four segmenters might be needed. The flexibility provided by the IRIS Explorer system made it possible to test many versions of our vision system with dispatch.

A second navigator module (*Navigator 2*) then provides graphic displays of the SWIMMER's behavior as it navigates to and then "devours" the appropriate target objects or avoids the inappropriate ones. Although this visualization is progressing, the shape recognizer (*ShapeRec*) module is classifying the object's shape. This information also contributes to the decision algorithms that guide and control the behavior of the SWIMMER. Not only does each representation of a module on the IRIS Explorer map serve to identify and logically locate the particular function executed, but it may also contain controls that allow the programmer to manipulate various parameters of the module's function. This is accomplished by means of simulated dials and buttons (called "widgets" in the IRIS Explorer terminology) that can also be varied by means of the computer mouse. Figure 4.3 expands the view of the color, intensity, and texture modules

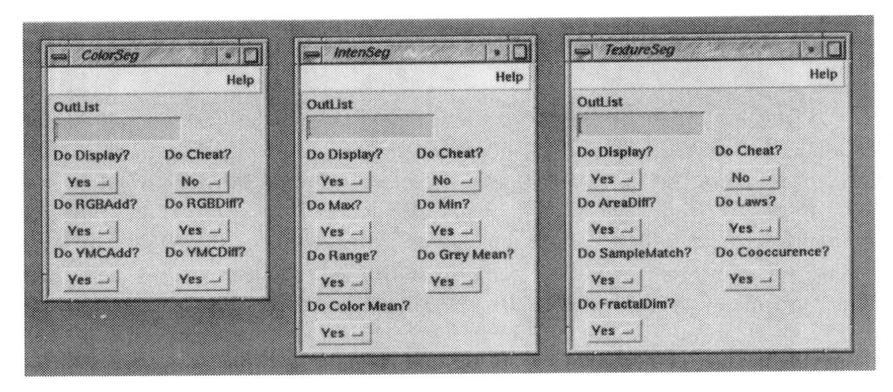

Figure 4.3 The buttons and "widgets" available on the color, intensity, and texture modules to control various aspects of their performance.

to show some of the control buttons. Using this feature, computer experiments can be carried out with ease. For example, manipulating the appropriate buttons on the *TextureSeg* module allows the experimenter to select which of the five available texture analyzer outputs are to be combined in any given experiment. In this way, a variety of experiments can be carried without any detailed source code programming of the system itself.[1]

4.4 VISUALIZATION

Because of the complex nature of the vision system and the need to observe its operation, we provided two additional ways for the experimenter to track its performance. Not only did this add to our intuitive understanding of the program's operations as it progressed, but it also was a powerful debugging tool in the event something did not function correctly. These two additional visualization modes tracked the course of the computer program and the behavior of the SWIMMER vision system simulation, respectively.

4.4.1 Visualization of the Computational Process

Figure 4.4 is a black-and-white snapshot showing one step of the model's operation. In this figure the several panels show the steps in the intensity, color, texture, and motion segmenters and their respective outputs taking place at this moment. In a subsequent frame, the process of combining is shown step by step in the successive frame of another panel. The advantage of this type of ongoing display is obvious: It permits the experimenter to observe the ongoing operation of the vision system. This is a powerful means of identifying any difficulties that may occur as the computational process proceeds.

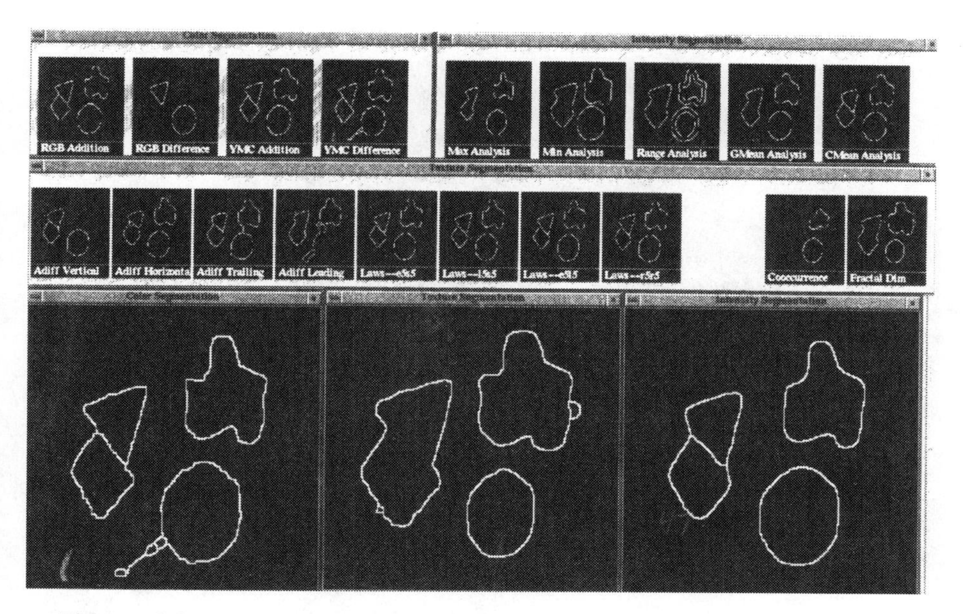

Figure 4.4 A snapshot of the dynamic display showing the momentary state of the operation of the model. Intermediate results from the color, intensity, and texture algorithms and the cumulative outputs of each set of algorithms are shown below.

4.4.2 Visualization of the System Performance

Figure 4.5 is a black-and-white snapshot of the dynamic display of the behavior of the SWIMMER model. The full motion picture series from which this single frame is taken is a presentation of the course of the entire process of the full model, in terms of the simulated behavior rather than the computational steps. An underwater "world" has been defined and stored in the computer memory. As the SWIMMER maneuvers to pursue "food objects," the background scene changes appropriately.

Since we did not have a real ocean and a real submarine with which to collect images and in which to observe behavior for most of the later stages of the project, we set up a "dry" virtual ocean in our laboratory and physically placed objects of various kinds in it. This "dry" scene information was acquired by our cameras and incorporated into the model's database—its virtual world. This information was made available to the vision system as if it had been acquired at the exact moment a real submersible would have observed the objects in a real ocean. In most simulations the objects' locations in the world model were determined randomly. This means that the simulated behavior in each experimental run was different. Simple rules governed the behavior:

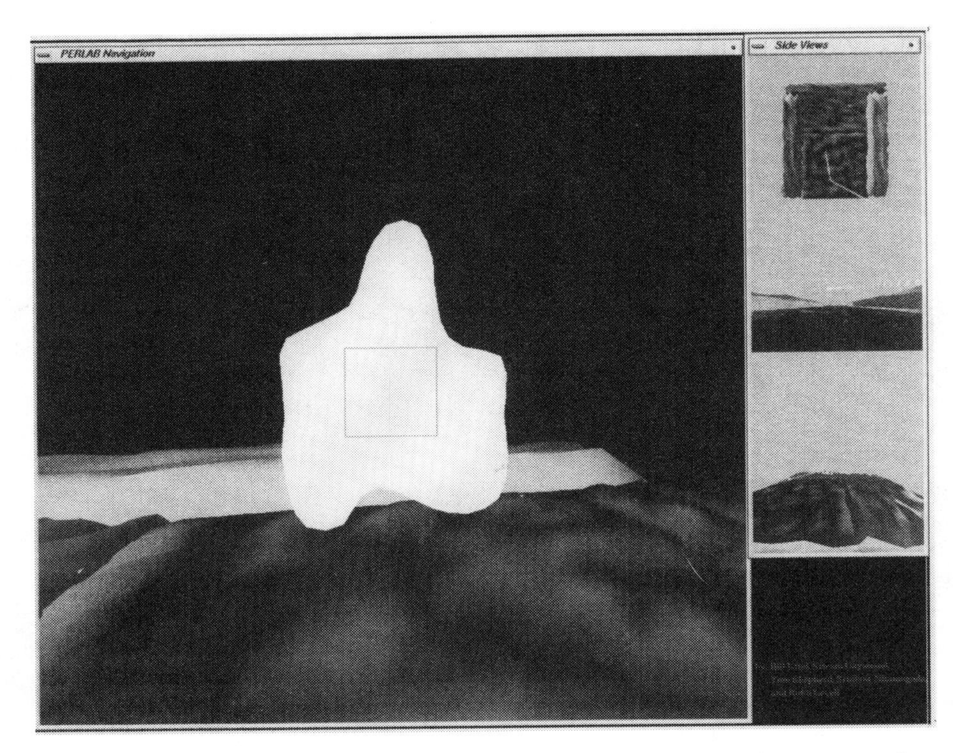

Figure 4.5 The behavior of the SWIMMER is visualized in a motion picture as the system processes the various outputs. This is one frame from the motion picture. It shows the SWIMMER's-eye view in the large frame of one of the targets in the simulation. The three smaller frames show "God's-eye" views from the top, side, and oblique viewpoints, respectively. The solid square represents a target that has not yet been recognized. The outline squares are aiming and trajectory marking aides.

Go to the nearest object.
Swim through the regular-shape objects.
Swim around the irregular-shape objects.
Stop when there are no more objects in the scene to be processed.

The visualization of the SWIMMER'S behavior shown in Fig. 4.5 consists of four frames. The first is the large frame on the left that represents the SWIMMER's-eye view—i.e., what the SWIMMER sees. The three smaller frames on the right side of this figure are "God's-eye" views of the entire scene as if viewed from the top, from an oblique angle from the side, and from an oblique angle from the back. The history of the navigational track of the SWIMMER is plotted by the white lines in these three smaller "God's-eye" view frames.

In the large SWIMMER's-eye view frame, white squares appear when an object is initially detected. These square shapes represent the SWIMMER'S awareness that an object is present (perhaps corresponding to accepting a SONAR signal as an indication of a detected target), but one whose shape has not yet been determined. The two squares are navigational guides showing the path that must be followed to approach the target object so that it will be in a frontoparallel orientation when the SWIMMER makes contact with it. The navigational aspects of the SWIMMER have already been described in Uttal et al. (1992) and need not be repeated here. It should not be overlooked, however, that this was a major programming effort in its own right.

At an appropriate distance between the SWIMMER and the object (determined by a single threshold parameter in the program), the shape of the object is provided by the segmenter to the simulation. This corresponds to the addition of short-range video information to the long-range, but shape-insensitive, SONAR information previously mentioned. At this point, the pattern recognizer (described in Chapter 7) is invoked to analyze and classify the object's shape. Depending on whether that shape is regular or irregular, the SWIMMER follows the rules described earlier.

With this general description of the function of the SWIMMER, we can now turn to detailed discussion of some of the more important modules that make up the complete vision system.

NOTE

1. The "*Do Cheat*?" button allowed the program to bypass actual execution of this step and to use a previously executed calculation for demonstration purposes.

5

A Particle System Model for Combining Edge Information from Multiple Segmentation Modules[1,2]

5.1. INTRODUCTION

So far in this book we have described how the prevailing school of thought in vision theory proposes that perceptual information representing several stimulus attributes, e.g., color, texture, motion, and depth, is transmitted along separate but parallel channels and then combined at some central location to provide a complete perceptual experience. In this chapter we present a method for combining information from multiple segmentation modules to obtain estimates of the most plausible edges in an image. This new method, which we call the *particle system* model, follows up on the statistical *center of gravity* model, in which spatial averaging was used. The statistical method, previously described in Lovell et al. (1992) and in the predecessor to this book (Uttal et al., 1992), produced a robust estimate of the edge of an object in the visual field by simply taking the spatial average of what were fallible and incomplete estimates produced by as many as 25 texture-sensitive discriminators. This part of our project, therefore, contributes to the segmentation portion of the model described in Chapter 4.

Segmentation, the extraction of regions and their bounding edges, has been a central task in the field of computer vision. Successful object segmentation is a basic and critical part of many applied fields, including robotics, medicine, and satellite technology, among many others. The process is also known as *figure–ground separation* in studies of human visual perception. The work we describe in this chapter, therefore, is a central part of our attempt to model human vision and to develop a powerful computer vision system analog for it. There is, perhaps, nowhere else in our model where the advantages of combination are so clearly demonstrated. The individual algorithms are all too frequently failures in demarcating the boundaries of regions. Even those algorithms that sometimes achieve an approximately successful specification of the boundaries in one image do not usually generalize; virtually any texture discriminator must ultimately fail in some situation where a different kind of texture cue is present.

Objects present in a scene can vary significantly in terms of their other attributes, such as intensity, color, depth, and movement. It is not a priori possible to specify which of the characteristic attributes will be present in any given scene or which will be the most salient. Ideal situations rarely occur in natural images, and the presence of sufficient and appropriate cues for any particular segmentation scheme to succeed is not always guaranteed. In spite of the fact that this was appreciated quite early in the history of the computer vision field (e.g., by Fram and Deutsch, 1975), until recently most efforts to develop segmentation algorithms have been directed at the fine-tuning of individual segmentation algorithms.

Along with an increasing number of contemporary workers in this field, we have argued in this book that the vigorous pursuit of an individual algorithm sensitive to only one of the possible attributes present in an image is not likely to be the best strategy for image segmentation. Such a "refine and polish" approach requires enormous potency on the part of the individual algorithm so that it can handle many different kinds of images. It is now clear that such an approach cannot be successfully implemented to operate in most real-world situations. First, the amount of control we have over the environment may not be sufficient to reduce all images to a standard form suitable for even the most polished and refined individual algorithm. Second, dynamically changing environments pose a severe problem, since image quality and the characteristic properties of an image may vary drastically. These difficulties can a priori render impossible any attempt to fine-tune an individual algorithm to process all potential images satisfactorily.

The preferred approach to which we are committed is to employ a set of algorithms or techniques each of which may be sensitive to a specific set of attributes and insensitive to others and then to combine their outputs. Even though no single algorithm may be individually robust or universally capable, the strategy of combination that we have championed here is also the one, as discussed in Chapter 2, that has evolved in the organic nervous system. We believe that in the long run, combination is probably going to be the only effective way to meet the challenge of approximating human vision with a computer. Both biology and computer science now attest to the fact that a successful strategy will have to involve appropriate combinations of the outputs of several algorithms, however frail and inadequate they may be individually, to obtain a robust final result.

A major advantage of the combination approach is that it is able to handle spurious responses from algorithms operating on images that themselves do not provide sufficient cues for their accurate performance. The major disadvantage of the combination approach is that it is computationally inefficient. That is, many different algorithms may have to be evaluated, since we do not know a priori which ones will be appropriate. However, even this disadvantage may be mitigated. Clearwater et al. (1991) have suggested that there is a fundamental advantage—the individual programs to be combined may be simpler than monolithic ones—to be obtained by the very act of cooperation in computational models over individual algorithms. They say:

> This work suggests an alternative to the current mode of constructing task specific computer programs that deal with constraint satisfaction problems. Rather than spending all the effort in developing a monolithic program or perfect heuristic, it may be better to have a set of relatively simple cooperating processes work concurrently on a problem while communicating their partial results. (p. 1183)

It is just now becoming widely appreciated that their admonition is more generally valid—it is solely by means of combination or cooperative techniques that any high-quality segmentation will ultimately be achieved. In that sense, it is clear that any initial fear of computational inefficiency would be short-sighted.

It is also increasingly apparent that data fusion or combination techniques are gradually becoming more popular in the literature of computer vision. Abidi and Gonzalez (1992), Aloimonos and Shulman (1989) and Clark and Yuille (1990), for example, have all reviewed the increasing importance of these techniques in recent years, as we have in Chapter 3. In addition to the topics we discussed there, researchers have also used the concepts of regularization theory (Poggio et al. 1985) and Kalman filtering (Matthies et al. 1988) to combine visual information. Following the taxonomy of Clark and Yuille, the particle system technique we propose here can be classified as a Class III, weakly coupled system. Such a system is characterized by two properties: (1) All of the individual modules act independent of each other; and (2) all of the data are not absolutely necessary to produce a final solution.

The combination technique we now propose is loosely based on the concept of particle systems first proposed by Reeves (1983) for rendering dynamic scenes in computer graphics. His inspiration, of course, was drawn from gravitational physics. This new procedure is more intuitive in a physical sense than the spatial averaging technique our group previously developed (Lovell et al. 1992). In addition, it integrates some of the tasks preceding and following the actual combination process, such as region extraction, edge thinning, removal of "twigs," and closing small gaps in edges in the course of its action. The method dynamically generates the objective edges of the image. This is similar to the work of Kass et al. (1987) and Durbin and Willshaw (1987), where the edge was initially estimated and then allowed to deform into the final required result. The next section gives a brief description of the concept of particle systems in general. A detailed description of the specific combination method we propose is presented in subsequent sections of this chapter.

5.2 THE PARTICLE SYSTEM APPROACH

Particle systems were first proposed as a modeling technique in computer graphics for objects whose behavior over time cannot be described in terms of their instantaneous spatial characteristics, as described by Foley et al. (1990). Reeves (1983) first used this technique to model structures like fire, explosions, and smoke. A particle system is defined as a collection of objects: the particles. The behavior of these particles is controlled by defining probabilistic rules. These rules control attributes of the system, like the generation of new particles, the extinction of particles, and their motion, color, and intensity characteristics. A typical particle entity P_i in a system like this can he described as a tuple of the following form:

$$P_i = (c_i, \mathbf{p}_i, m_i, \mathbf{v}_i, l_i)$$

where

c_I = color of the particle
$\mathbf{p}_i$ = current position of the particle
m_i = mass of the particle
$\mathbf{v}_i$ = current velocity of the particle
l_i = specified lifetime of the particle

Rules of generation, extinction, and motion govern the behavior of the particle system over a period of time. These rules specify the conditions under which new particles are generated and terminated (depending on the lifetime l_i of the particle) and also the laws of motion that the particles follow to generate the required trajectories. It should also be appreciated that the particle attributes that we have just listed are not the only ones possible. Different models can involve other relevant attributes that control the overall behavior of the system of particles.

Another important part of any particle system model is a rendering scheme to translate the complex behavior of the system of particles into a viewable and easily understood picture for the user. Such a rendering scheme has been incorporated into our model, as described later.

5.3 A PARTICLE MODEL FOR IMAGE SEGMENTATION

The rationale behind the particle system we propose in our model is similar to that motivating the general case just described. The main advance that we have made is the definition of the specific rules governing the existence and behavior of a typical particle in the system. These rules provide answers to the following detailed questions governing the behavior of the system: (1) When will a new particle be generated? (2) Where will it be placed? (3) What forces will act on a particle to determine its motion? (4) When and where will an existing particle be removed? (5) When should the process terminate?

In a computer simulation of a natural system, the real physics of the phenomenon being modeled provides the answers to these questions. For example, a system modeling the behavior of smoke can use entropy as a determinant to control the generation and extinction of the constituent particles. The generation of new particles can be controlled by probabilistically characterizing the source of the smoke to account for variation in the intensity of emission. The motion of the smoke particles is directly controlled by factors such as air resistance and wind velocity that are introduced in the model. These factors are always particular to the phenomenon being modeled and can encompass a wide spectrum of physical properties for different applications.

In the model we propose, the objective is to fuse or combine the edge estimates emerging from a set of independent segmentation modules into a single robust "best estimate" of the edge. To make this objective clear, consider the example shown in Fig. 5.1. Figure 5.1A shows a sample image of a collage of pieces of different materials of varying texture, color (in the original), and intensity. Figure 5.1B shows the output of several segmentation methods operating on the image. The segmentation techniques that produce each of the individual estimates of the boundaries are of several different kinds. These include the area differential in four principal directions, the fractal dimension, the co-occurrence matrix, Laws' microtexture masks for texture; the maximum, minimum, and range for intensity; and finally an opponent and cooperative color scheme. These procedures are discussed in detail in an earlier paper from our laboratory (Lovell et al., 1992). Three of them—Laws', fractal, and co-occurrence—are standard methods previously published. The others are novel to our laboratory and exhibit sensitivities to different attributes of the input image than the standard ones.

It is obvious that the outputs of the various methods vary significantly, with some performing significantly better than others. The new fusion technique that we have based on the concept of particle systems is designed to operate on the

Figure 5.1 (a) A test image of four pieces of carpet material of different color and texture against a clear white background. (b) The outputs of 15 different segmentation techniques sensitive to texture (area differentials, fractal, laws, and co-occurrence), intensity, (max, min, and range), and color (RGBdiff and RGBadd), respectively. It can be seen that the performance of these methods varies significantly.

aggregate of the outputs of these individual algorithms to produce a single best estimate of the edges in the image.

There is one major problem faced from the outset in our application of particle system theory. In the case of computer image processing, there is no particular physical phenomenon being modeled. Rather, we have had to invent a system of rules governing the existence and motion of particles to respond to the spatial locations of the edge estimates from the segmentation modules. To do so, the outputs of the set of segmentation methods are interpreted as a force field, with the edge information produced by each of the individual segmentation algorithms producing hypothetical lines of force. Our trick is now to introduce an

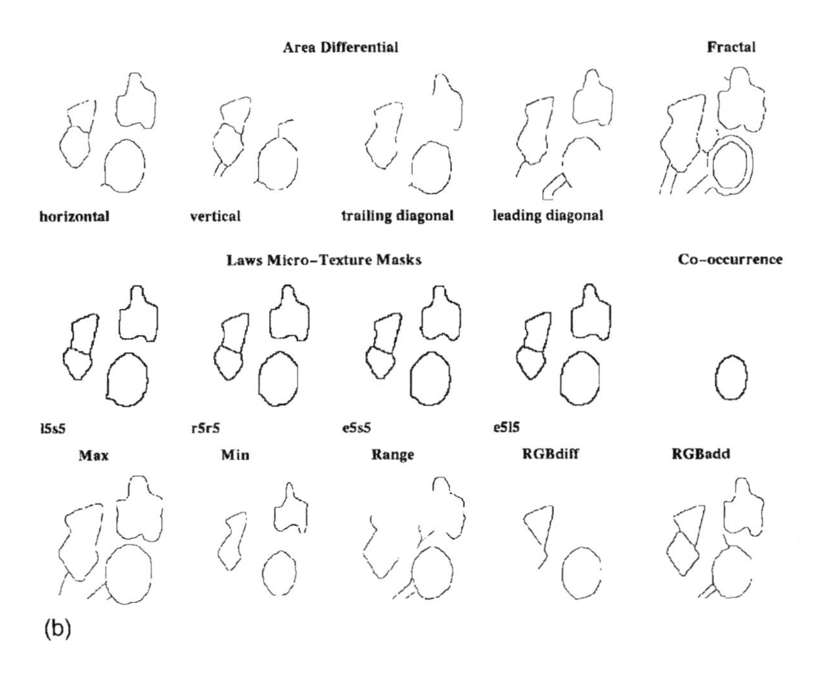

(b)

additional simulated probe particle into this force field and to observe its motion as it travels along under the influence of the lines of force. Our argument is that the trajectory of the introduced simulated particle (modified by some constraints that are described later) will produce an excellent estimate of the edges of the objects in the image.

All of the outputs of all of the segmentation models are superimposed into a single space. This produces the rather noisy and overlapping picture shown in Fig. 5.5A. The system is then initialized by inserting the probe particle at a location selected because it has the highest density of edge information within any local area of this cluttered picture. Additional probe particle generation will occur only in the event of a bifurcation in the path of the particle or when the particle has traversed a closed path.

The forces causing a probe particle to move from its current position are analogous to the Newtonian forces of attraction between that particle (at a particular location) and all of the edge information in a region of fixed size around it. Due to the discrete nature of digital images, the edge information itself must viewed as a collection of discrete elements located at identifiable (x,y) pixels in the image plane. It is these individual discrete edge elements that exert the forces of attraction on a particle.

The probe particles that are introduced to track the influence of these edge information pixels are removed at a later stage when they have fulfilled their function of establishing the image edges. When all the particles in the system reach an equilibrium or quiescent state, their recorded trajectories or paths are considered to be robust estimates of the image edges. The test for the equilibrium state will also be used to terminate system execution. These and other factors characterizing the particles, the forces acting on them, and their behavior are described more formally in the next section.

In addition to being intuitively satisfying, since the system is modeled in the context of observable and easily conceptualized physical forces, this approach has some other advantages. Observing the behavior of the particles (using computer graphics techniques) in real time makes it possible to observe directly the performance of the system as it progresses. However, it is important to remind our readers that that the model we propose is not meant to describe specific neurophysiological mechanisms in the human visual system. It is a simulated functional or behavioral analog capable of carrying out a task that we know the human visual system performs with relative ease—combining multiattribute edge information. Furthermore, we believe it to be sensitive to some of the same variables as is human vision and to embody the idea of fusion or combination in a way that is analogous, but not homologous, to the mechanisms present in the organic vision system. But obviously, it is not likely that the forces of gravitational attraction are the ones at work in human vision. Whatever forces exist there are currently unknown, and our model only describes, but does not reductionistically explain, the system's function. That is all that can be expected from a mathematical or computational model.

Now let's discuss the specific details of the model. Henceforth, the edge information (i.e., the spatial location of the pixels on the edges) obtained from the segmentation modules is referred to as the collection of data elements. Each data element is nothing but a point (due to the discrete nature of digital images) in the coordinate system employed by the particle system. The location of a data element is denoted by $\mathbf{d}_{ij}$. A particle location is denoted by $\mathbf{p}_i$.

A typical particle P_i in our model can be characterized by a tuple

$$P_i = (\mathbf{p}_i, \mathbf{s}_i, \mathbf{v}_i, s_i, \delta, e_i, c_i, l_i),$$

where

$\mathbf{p}_i$ = current position of the particle
$\mathbf{s}_i$ = point at which the particle was spawned
$\mathbf{v}_i$ = current orientation of the particle
s_i = size of the region of influence around P_i
δ = magnitude of the advancement of the particle
e_i = number of endpoints encountered
c_i = number of closed loops executed
l_i = lifetime of the particle

The particle system we have employed is a homogeneous one. In other words, all probe particles are identical in every aspect of their generation, extinction, and motion characteristics. We now describe the just-mentioned parameters in more detail.

5.3.1 Current Position and Orientation

For a particle P_i, $\mathbf{p}_i$ specifies its position in the image plane and $\mathbf{v}_i$ its current direction of motion. The position is specified in a two-dimensional Euclidean coordinate system, with the origin at the lower left corner of the image. The direction of motion is a vector in the same coordinate system that specifies the direction in which particle P_i approached its current position.

5.3.2 Spawn Point

For each particle P_i, the coordinates of the point $\mathbf{s}_i$ at which it was spawned are recorded. These are used to prevent repeated spawnings of particles in the same direction.

5.3.3 Region of Influence

Each particle P_i has a specified region of size s_i around its current location called its *local region of influence*. The edge information present within its region of influence is the only data that have any effect on the particle's behavior. This region is a square, and two of its parallel sides are oriented in the direction of the particle's current direction vector $\mathbf{v}_i$. The size of the square region is fixed for all particles in the system. The value of s is selected to span the spatial variation in the location of an edge element detected by different segmentation methods.

Figure 5.2 shows the interior of a probe particle's region of influence in detail. The square region is divided into a fixed number of slabs L, as shown. At any given position, the upper half of a particle's region of influence is of primary interest. The lower portion is used only to verify the direction of the probe particle's approach to a position $\mathbf{p}_i$ in the event of multiple bifurcations. At a particular position, each particle is assumed to be attracted by the data elements in the upper half of the region of influence. This force of attraction $\mathbf{F}$ is assumed to be the standard Newtonian force of attraction between two masses:

$$\mathbf{F} \propto \frac{m_1 m_2}{r^2}$$

We now describe in detail the computation of the new direction of motion for a particle P_i. At a particular position $\mathbf{p}_i$, a search is made of the slabs in the

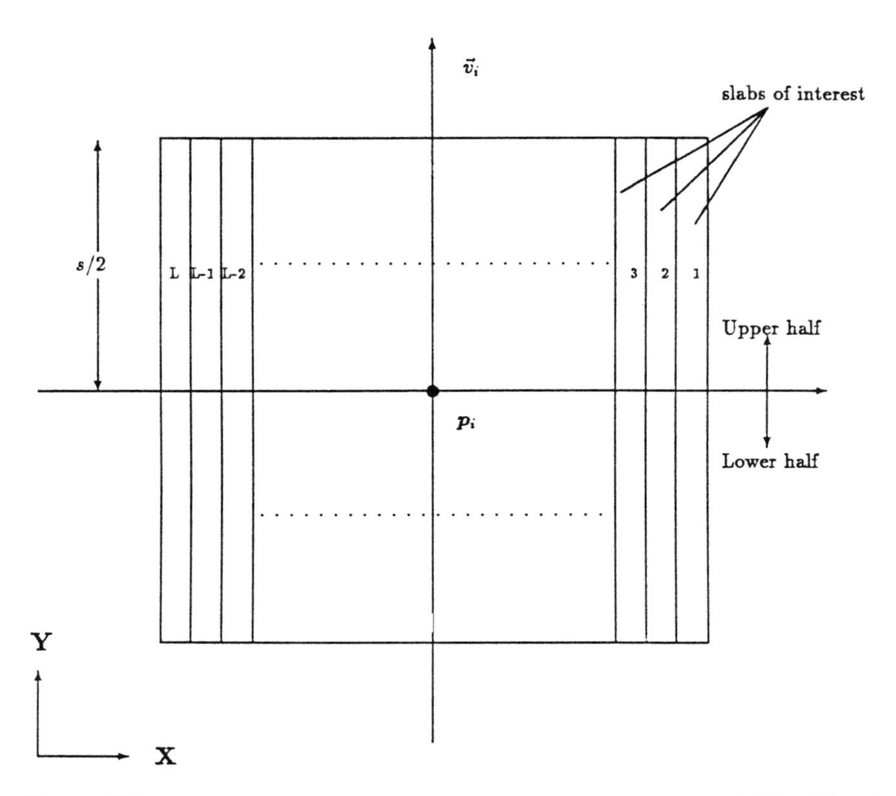

Figure 5.2 The region of influence of a typical particle in the model. The slabs of interest and the division of the region into two halves based on the current orientation of the particle are shown.

upper half of P_i's region of influence. The data elements present in each slab are detected. Let n_j be the number of particles in the jth slab. For each data element k in a slab j at position $\mathbf{d}_{jk}$, the vector between $\mathbf{p}_i$ and $\mathbf{d}_{jk}$ *is* formed. This vector is then scaled by the force of attraction $\mathbf{F}_{jk}$ between the particle and the data element. A weighted convex combination of these scaled vectors is taken to give us a resultant direction of attraction. Let us call this new direction of attraction $\mathbf{u}_i$. The particles in each slab j are weighted in an inverse manner to the slab's distance from $\mathbf{p}_i$. We have

$$\mathbf{u}_i = \sum_{j=1}^{L} W_j \sum_{k=1}^{n_j} \mathbf{F}_{jk}(\mathbf{d}_{jk} - \mathbf{p}_i)$$

where

$\mathbf{u}_i$ = new direction of motion
$\mathbf{F}_{jk}$ = force of attraction between the data element at $\mathbf{d}_{jk}$ and the probe particle
w_j = weight factor for particles in slab j
$\mathbf{d}_{jk}$ = position of the kth data element in slab j

The weight w_j for particles in slab j is selected in such a way that slabs closer to the center of the region of influence have more weight than the slabs closer to the boundary of the region of influence. This keeps the particle from drifting toward the extremities of the collection of edges. To reduce the effect of noisy data on the particle trajectory, the final estimate of the new direction of motion is taken to be a weighted sum of the current direction $\mathbf{v}_i$ and the vector $\mathbf{u}_i$ estimated in the manner just described. Let $\mathbf{z}_i$ be this vector,

$$\mathbf{z}_i = \alpha \mathbf{v}_i + \beta \mathbf{u}_i$$

where α and β are scalars and $\alpha + \beta = 1$. We used values of $\alpha = 0.2$ and $\beta = 0.8$. The effect of this weighted combination is that the particle offers resistance to a sudden and drastic change in its path of motion. The mass of all data elements and particles is assumed to be unity. This parameter can be used to weigh the influence of edge information from selected modules (when such a priori information is available) by increasing the mass of the data elements in their output.

5.3.4 Particle Motion

At every stage of the iterative computation process, each particle P_i in the system is advanced by a fixed amount δ in the new direction of attraction $\mathbf{z}_i$. After this motion, the current position $\mathbf{p}_i$ of the particle P_i is updated. Also, the new direction of motion $\mathbf{z}_i$ becomes the current orientation vector $\mathbf{v}_i$. δ, the fixed amount by which the particle is moved along the orientation vector, is regulated by the resolution of the image from which the edge data are obtained. Since the particles operate in a coordinate system of their own, this value is specified so as to be within the minimum distance possible between two data elements.

5.3.5 Particle Spawning

Bifurcations in the edge information present in the image are handled by spawning new particles to explore new branches. At each bifurcation point, the model ensures that one new particle is spawned in each of the directions of bifurcation. These newly spawned particles now become incorporated into the system and are processed at each iteration of the combination process. The success of this technique depends on the accuracy with which bifurcation points are detected.

n_p denotes the number of active particles in the system at every point in the iterative process.

After a new direction of motion $\mathbf{z}_i$ has been detected, the angle between $\mathbf{z}_i$ and the current orientation vector $\mathbf{v}_i$ is computed. An angle of 0° implies absolutely no change in the direction of motion, requiring no new particle spawning. At bifurcation points $\mathbf{z}_i$ diverges considerably from $\mathbf{v}_i$ and signals a possible bifurcation. In this event, the particle performs a "lookaround" operation to detect the directions in which to spawn new particles. This involves rotating the square region of influence around the current position $\mathbf{p}_i$ to span a 180° angle in a fixed number of steps and computing the direction of attraction $\mathbf{u}_i$ as before. Particles are spawned along directions where the newly detected values of $\mathbf{u}_i$ span angles greater than a threshold value with the current orientation $\mathbf{v}_i$.

In the ideal case, the correct number of particles are spawned at the appropriate locations and no spurious bifurcation points are detected. Since we cannot ensure that this will be the case for all possible data, it is better to follow a conservative approach and spawn particles even when we are not completely sure about the existence of an actual bifurcation point. The process is self-correcting, and no damage is done by spawning extraneous particles. Mechanisms are inherent in the process that detect these "useless" particles and terminate their existence. These mechanisms are discussed later. This approach, though it increases the amount of computation in the worst case, ensures that we do not entirely miss any edges.

5.3.6 Closed Paths and Endpoints

Edges in the image can be of two types: closed and open-ended. The system is designed to handle both types. For each particle P_i, a track is kept of the number of endpoints (e_i) encountered. Endpoints are detected when the number of data elements in a particle's region of influence approaches zero. When this occurs, the current direction of the particle is reversed (rotated by 180°) and the particle traces its path backwards. Closed paths are detected by keeping track of the point of conception of each particle. When a particle's trajectory brings it back (within a tolerance) to this point, a closed path is assumed to exist. c_i denotes the number of closed loops a particle executes. These values are used to terminate system execution, as explained, at a later stage.

5.3.7 Particle Lifetime

For each particle, the lifetime l_i denotes the number of iterations for which the particle has been in existence. In other words, all particles start with l_i having an initial value of 0. Each time the particle advances along its new direction of motion $\mathbf{z}_i$, the value of l_i is incremented. This value is used in conjunction with

e_i and c_i to terminate system execution. It also aids in the important task of terminating spurious particles.

5.4 THE ALGORITHM

This section describes the iterative algorithm that performs the actual fusion process. The algorithm is guaranteed to terminate, and the final output trajectory is a single, fused representation of the estimated edges in the image. We concentrate on describing the two major aspects of the algorithm, the iterative part and the test for quiescence, which is used to terminate the execution of the iterative process.

5.4.1 The Iterative Part

The iterative part of the algorithm commences with the insertion of the first particle into the image consisting of the data elements. This initial particle is positioned by scanning across the input image and detecting the region of highest data element density. The size of the local region used is the same as the size s of the particle's region of influence, as we described earlier.

The initial value of $\mathbf{v}_i$, the current orientation vector, is computed at this point by rotating the region of influence in discrete steps through a 360° angle and by determining the direction in which the maximum number of data elements is encountered. The parameters s and δ are fixed based on the image resolution and are constant for all particles in the system.

The outline of the algorithm is shown in Fig. 5.3. The various steps involved in the algorithm FusionIterate have been described in detail earlier when we described the parameters of a typical particle in the system. It can be seen that the iterative loop consists of a very simple sequence of steps. This is possible because much of the detail is encapsulated in the description of the behavior of the particles in the system. Also, the particles we utilize in the system are homogeneous; i.e., all possess identical characteristics and sensitivities to factors determining their behavior. These two factors in turn simplify the computation to a great extent.

In addition to the primary task of detecting the edges in the image, the algorithm also performs region extraction, as defined by Rosenfeld and Kak (1982), of closed areas in the image. The regions are extracted by a labeling operation that is built into the probe particles' behavior. All data elements are assigned an initial label value of −1. For each iteration of the algorithm starting with the detection of the densest location, a count of the number of regions (called the *region count* and denoted as ρ) detected up to that point is incremented. In the computation of the new direction of motion $\mathbf{z}_i$ of a particle P_i, the data elements considered are only those with a label of −1 (data elements that have not been

algorithm FusionIterate

- *Insert initial particle into remaining unlabelled data. Set* $n_p = 1$.
- **while** (*NOT* **Quisence**)

  for *each particle* P_i, ($i = 1$ *to* n_p) {

  1. *Compute new direction of motion* $\vec{n}_i$.
  2. *Compare* $\vec{n}_i$ *and* $\vec{v}_i$.
  3. **If** *possible spawn point, look around, and spawn particles in promising directions.*
  4. *Initialize* $\vec{v}_i$, $\mathbf{p}_i$, $\mathbf{s}_i$ *for all the new particles spawned.*
  5. *Advance particle* P_i *in the direction* $\vec{n}_i$ *by amount* δ *to* $\mathbf{p}_i$.
  6. **If** *endpoint, increment* l_i, *reverse direction of motion of* P_i.

     else if *closed path detected, increment* c_i.

  }

  Update particle count n_p.
- *Save trajectory of particles in the system at quiscence.*
- **If** *unlabelled data exists, set* $n_p = 0$. *Execute algorithm* **FusionIterate** *again.*

  else *TERMINATE.*

Figure 5.3 Algorithm FusionIterate.

encountered so far) or those whose label equals the current region count value ρ. New data elements encountered in this manner are then labeled ρ. If, at quiescence, a closed path has been detected, the edge produced is the boundary of the region detected. In the event of open-ended edges, the region labels are ignored at quiescence and only the detected edges are output. The algorithm then proceeds by detecting the densest spot in the remaining unlabeled data elements, increments the region count, and proceeds as before. This also ensures that all the data elements get processed and that the algorithm terminates when no more unlabeled data exist.

The test for quiescence is performed at the beginning of each loop of the algorithm FusionIterate. This test is described in detail in the next section. When this test succeeds, the iteration is suspended and the trajectories of the particles present are stored. The algorithm commences again if there are data remaining that have not been processed yet, i.e., data elements with a label of −1. When all the data have been processed in this manner, the algorithm terminates. The edges saved up to that point are the "fused" edge estimates.

5.4.2 The Test for Quiescence

An important part of the algorithm is the check for quiescence, as described earlier. The algorithm FusionQuiescence shown in Fig. 5.4 performs this task. The primary function of the algorithm is to check for the presence of closed paths and the existence of open-ended edges.

A closed path is characterized by a particle's looping through its path two times; i.e., $c_i \geq 2$. The second pass is performed as a double-check for the validity of the closed path generated during the first pass. Open-ended edges are characterized by the particle's visiting one of the endpoints twice; i.e., $e_i > 2$.

algorithm FusionQuiscence

- *Set* **closed_paths** = *FALSE and* **open_contours** = *FALSE.*

- **Check for closed paths**

  if *a closed path exists for a particle P_i and all other P_j also have closed paths(i.e. $c_i \geq 2$ and $c_j \geq 2$)*

  1. *For all P_j, Check lifetime l_j and retain the ones whose lifetimes differ significantly from l_i.*
  2. *Kill all other particles.*
  3. *Set* **closed_paths** = *TRUE.*

  else if *a closed path exists for a particle P_i and an incomplete closed path exists for some P_j (i.e. $c_i > 2$ and $0 \leq c_j < 2$)*

  Set **closed_paths** = *FALSE.*

- **Check for open ended contours**

  if *all particles without any closed loops have open ended paths (i.e $e_i > 2$)*

  1. *Compare endpoints and retain one particle for each endpoint pair.*
  2. *Check lifetimes of remaining particles and kill those with lifetimes less than the threshold σ.*
  3. *Set* **open_contours** = *TRUE*

  else if *an incomplete open contour exists (i.e $0 \leq e_i \leq 2$)*

  Set **open_contours** = *FALSE.*

- **if** *(***closed_paths** = *TRUE and* **open_contours** = *TRUE)*

  *return(***Quiscence** = *TRUE)*

  else *return(***Quiscence** = *FALSE)*

Figure 5.4 Algorithm FusionQuiescence.

Once a closed path has been detected for a particle P_i, all the remaining particles P_j, if any, are also checked for closed paths. In the event that all the P_j already exhibit closed trajectories, their lifetimes l_j become the factors that regulate their further existence. Values of l_j very close to l_i, the lifetime of particle P_i, indicate that particle P_j is a multiple copy of P_i and is following the same path. Situations like this arise due to the spurious spawning of particles along a path. Such multiple-copy situations occur when spurious particles are spawned along a path. Such particles are terminated. The variable *closed_paths* is set to TRUE, indicating that all particles remaining are traversing valid closed paths in the input data. If particles that have executed their closed paths only once still exist, *closed_paths* is set to FALSE, indicating the uncertainty about the validity of their trajectories. Processing of such particles is continued in the iterative loop until they complete their closed trajectories for the second time.

The second stage of the algorithm verifies the validity of open-ended edges. As described earlier, a value greater than 2 for the parameter e_i indicates an open-ended path in the trajectory of a particle P_i. This check is limited to the particles that do not execute closed loops; i.e., $c_i = 0$. In the event that all such particles have detected open-ended paths, a check is made for multiple traces of the same path by more than one particle. This is performed by comparing the endpoints of their paths against those of the remaining particles. As with closed paths, this situation occurs due to spurious particle spawnings. These duplicate particles are also terminated.

As we noted earlier, the particle system in our model has a "detwigging" operation incorporated into the particles' behavior. This removes line segments with lengths less than a specified threshold value. This is performed by comparing the lifetime (which is just another measure of length) of particles with open-ended paths against a predetermined value. Particles whose lifetime falls below this value are terminated. After these operations, the value of the variable *open_edges* is set to TRUE, indicating that all particles remaining are valid open-ended edges in the input data. If there are particles that have not yet visited one of their endpoints at least twice, this variable is set to FALSE, indicating that these particles need further processing.

After the check for closed paths and open-ended edges has been performed, the status of the variables *closed_paths* and *open_edges* are checked. A value of TRUE for both indicates that all possible closed edges as well as all open-ended ones in the trajectories of the particles currently present have been detected. The algorithm then returns a value of TRUE to algorithm FusionIterate, which in turn stores the output of the edges. A value of FALSE for either variable indicates that further processing is necessary, and the algorithm returns a value of FALSE to FusionIterate.

The edges output at quiescence therefore track the trajectories of the valid particles and hence have minimal thickness when recorded in the image plane.

Figure 5.5 (a) The aggregate of the edges as detected by the segmentation routines on the image shown in Fig. 5.1a. This is formed by superimposing the outputs of the individual algorithms. (b) The fused edge estimates.

In other words, the model has an automatic thinning procedure built into it. Also, size thresholds can be used to eliminate open-ended edges and closed paths of size lesser than these minimum values. This can be done only when a priori knowledge of the nature of the edges expected from an image is present. This aids in eliminating regions and lines in the final estimate that are present due only to noise or spurious responses of the segmentation modules.

5.5 RESULTS

The fusion technique we have described in this chapter has been tested on a variety of images with varying values of the characteristic attributes: color, intensity, and texture. In this section we present the results of the performance of our method on three such images. All our test images had 24 bits of color

Figure 5.6 (a) An image of two overlapping pieces of carpet material of different colors and texture against a wood-grain background. (b) The aggregate of the edges detected by the segmentation methods. (c) The fused edge estimates from our model.

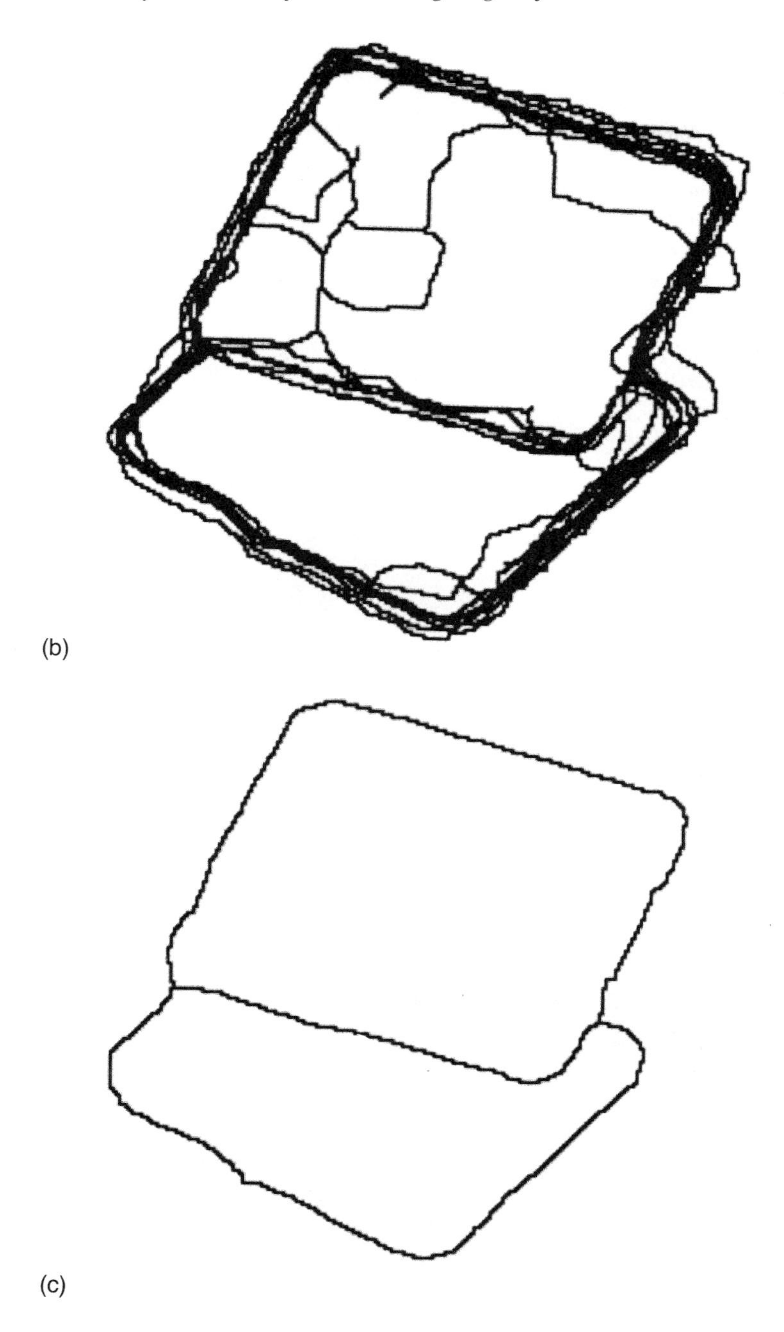

(b)

(c)

resolution (8 bits each of red, green, and blue) and were 256×256 pixels in spatial resolution. The segmentation methods used operated with the same parameters and thresholds on all the test images. It must be mentioned here that the proposed fusion technique does not in any way depend on any particular set of segmentation methods. On the other hand, the success of the method is tied to the ability of the segmentation methods to generate collectively as much of the edge information present in the image as possible.

The first test image (shown earlier in Fig. 5.1A) consists of four pieces of carpet material, of different types varying in their color and texture composition, over a clear white background. The outputs of the individual segmentation methods are superimposed to produce the aggregate image shown in Fig. 5.5A. Figure 5.5B shows the output of the fusion technique in which the aggregate image was reduced to a single edge.

Figure 5.7 (a) An image of an underwater scene with objects scattered on the base of a pool. (b) The aggregate of edges detected by the segmentation routines. (c) The fused edges.

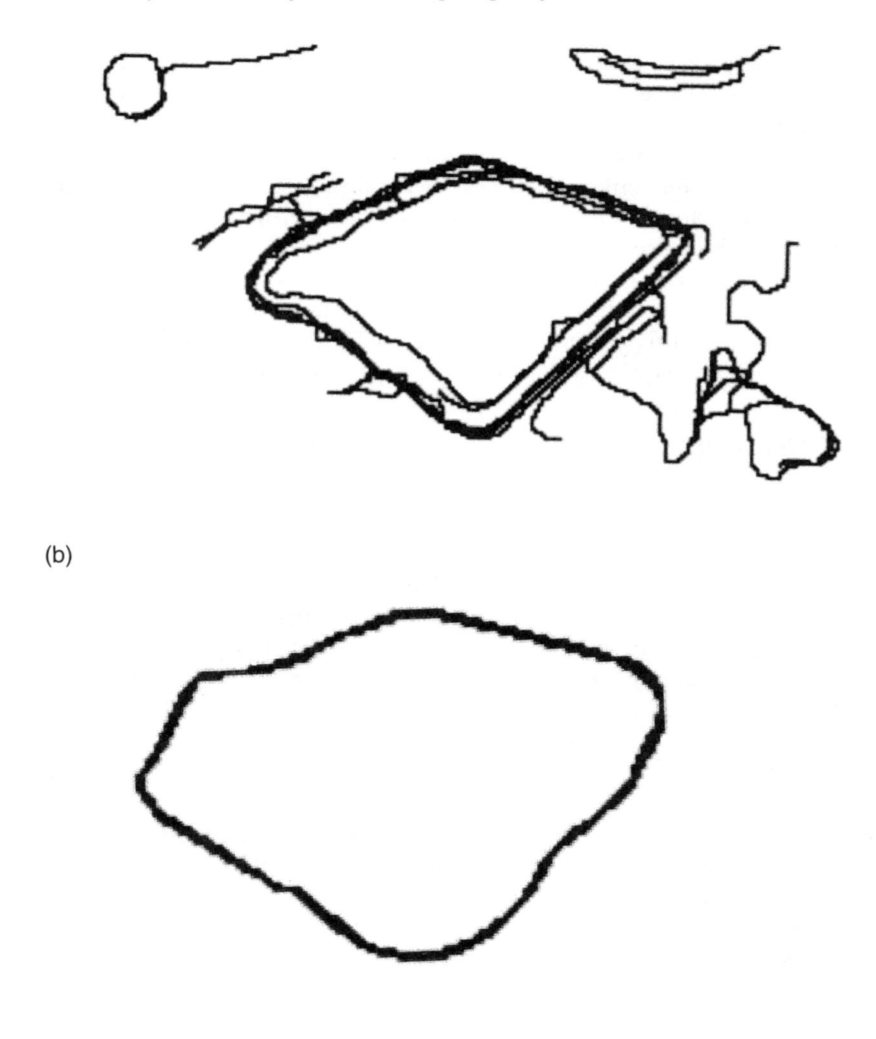

The second test image, shown in Fig. 5.6A, consists of two overlapping pieces of carpeting against a wood-grain background. Here the entire image possesses texture to some degree and at the same time has a variation in the intensity and color content also. Figure 5.6B is the aggregate of the edges detected. Finally, Figure 5.6C is the fused edge estimate.

The third and final test image (shown in Fig. 5.7A) is that of an underwater scene in a test pool with some objects scattered on the floor of the pool. This image is a very good example of a real-world situation with insufficient and imperfect lighting conditions. In addition, due to the diminished visibility under water, the image has a washed-out look, with surfaces appearing smoother than they actually are. Not surprisingly, most of the texture discriminators we employed did not perform well on this image. On the other hand, there are sufficient intensity differences (and, in the original image, color differences) for the segmenters that operate on intensity or color to detect successfully the outlines of the objects. These attributes produced the aggregate image (shown in Fig. 5.7B), which has sufficient information for our fusion technique to process. Figure 5.7C is the final representation of the edge of the object after fusion.

5.6 CONCLUSIONS

In this chapter, we have presented a technique to combine the outputs of various segmentation operators acting on a variety of image attributes into a single *fused* edge representation. We adopted the philosophy that combining the outputs of a set of segmentation operators is preferable to attempting to design a single operator that performs uniformly well over a wide variety of images. This approach was stimulated by new developments in perceptual psychology, neurophysiology, and anatomy, as discussed in earlier chapters. Our tests with a variety of natural images under varying lighting and visibility conditions have shown that with a judicious selection of operators sensitive to several image attributes, we can achieve a high degree of accuracy in image segmentation tasks, despite the inadequacies of the individual operators in dealing with particular images. In addition, the method has an intuitive physical interpretation that makes it easy to comprehend the mechanisms involved in the combination process. Finally, we reiterate that this method is not an attempt to model any specific mechanism in the human visual system. Rather, it is an analogous procedure for performing a perceptual task the human visual system executes with great ease.

NOTES

1. This chapter is adapted from Dayanand et al. (1994) and is used with the permission of the publishers.
2. We specially acknowledge the contributions to this paper of Robb Lovell, who was the author of the segmentation modules used to test our combination procedure.

6

Combining Images for Three-Dimensional Vision[1,2]

6.1 INTRODUCTION

So far in this book we have discussed the general idea of combination and given some examples of the progress that has been made mainly in combining two-dimensional images. There is another domain, however, in which the challenges of combination are even more extreme: three-dimensional vision. It is here that psychophysical evidence makes it abundantly clear that combination is a central part of three-dimensional information processing in the organic visual system. At the very least, the stereoscopic perception of depth and surface shape must arise as a result of the combined processing of the two two-dimensional images projected onto the retinas. However, many other monocular and binocular cues are also known to affect the outcome of this transformation. A number of these monocular and binocular cues have been identified, each of which is capable of contributing veridical spatial perceptions. Monocular cues include (1) interposition, (2) geometrical perspective, (3) areal perspective, (4) relative motion, (5) shading, and (6) accommodation, among others. Another binocular cue helpful in reconstructing the three-dimensional attributes of a scene is the proprioceptive awareness of the convergence angles of the lines of sight. Finally, the accommodative state of the eye's lens is also known to provide a compelling physiological depth cue.

It is in the process of creating depth from the invariant properties of two-dimensional images that the impact of higher-order knowledge (another form of information combination, but one that we do not deal with here) is most evident. As we see, many two-dimensional images that are the projections of solid objects just do not contain the information necessary for their reconversion into three-dimensional perceptual experiences. It is only by combining several cues, or adding information from previous experiences, that these "ill-posed" perceptual problems can be solved and a coherent perception arise. For example, the classic Neckar cube or the Schrödinger staircase illusions are both inadequately defined by two-dimensional stimuli. This ambiguity results in an uncertainty concerning which of two possible interpretations of the two-dimensional projection is correct. The perceptual response to the figure is, therefore, free to alternate or reverse periodically in accord with other cognitive influences that are only beginning to be understood. Sometimes the additional information can be obtained directly by analyzing the image itself in some other way. For example, adding information from the lines of projection in an image to the information from shading can resolve the ambiguities in interpreting shading information. This is the one of the directions we have taken and is prototypical of the work on visual combination that we present here.

The question of how the visual system reconstructs three-dimensional space from its two-dimensional inputs has been approached from a number of points of view. Psychophysical and neurophysiological studies have illuminated some of the functional and structural aspects of the problem, but the computational modeling approach offers a powerful additional means of seeking answers to this question. It does so by forcing precision in proposed models as well as by providing a means of reality-testing proposed theories. Most of all, however, computational modeling, drawing on its sister endeavor—computer vision—is itself a rich source of alternative theories and models of human three-dimensional space perception. Though currently challenging for computers, the natural "existence proof" represented by human vision is strong evidence that the reconstruction of the three-dimensional attributes of a scene from two-dimensional images can be successfully accomplished. It also helps to confirm an ideal goal toward which work in this field should be directed. A useful review of the basic human psychophysical data and related theories of three-dimensional vision can be found in our earlier work (Uttal, 1981).

In recent years, computational theories have emphasized three cues in particular for the reconstruction of surface shape and range when the input information is only two-dimensional. In the discussion in this chapter we concentrate on static snapshots. We do not deal with what may be one of the most vigorous fields of study concerning three-dimensional shape reconstruction—shape from motion.

The three cues of special current interest to us are (1) stereo disparities (a binocular cue), (2) shading, and (3) geometrical perspective (two monocular cues.) Unfortunately, as in the organic domain, none of these cues is completely satisfactory for depth reconstruction, for reasons that we individually describe later.

In this chapter we build on the two-dimensional parts of our model discussed in this book and its predecessor (Uttal et al., 1992) by reporting some techniques that we have developed to combine these three cues to recover three-dimensional range and surface shape from two-dimensional images. The present version of our three-dimensional combination model is based on improved and modified versions of three individual algorithms for stereoscopic vision (Stereo), shape from shading (SFS), and shape from structured light (SFSL) previously reported by other laboratories. We concentrate in this chapter on combination processes that lead to robust three-dimensional shape reconstruction. Our progress in combining these individual algorithms is presented as a process model of three-dimensional spatial perception by organisms. Although neither the specific computational algorithms nor the mechanisms they imply are likely to be those used by the brain, the general concept of combining, binding, integrating, or fusing different functions processing different cues helps us to understand how a collection of procedures can interact to improve the perceptual veridicality of visual space beyond that provided by any one alone.

The point of view that three-dimensional vision is mediated by an integrated system of independent modules has been expressed by a number of other contemporary authors. Bülthoff and Mallot (1990) describe a computational model of how disparity, shading, and texture cues can be combined to produce improved three-dimensional perception. Dillon and Caelli (1992) describe a hybrid system integrating stereo and focus cues. Jeong and Kim (1992) propose a unified theory of early vision in which optical flow, stereo disparity, shading, and texture cues are used collectively to reconstruct surfaces. As is shown later, our approach differs in both technique and philosophy from each of theirs.

There are several mathematical and computational arguments that any three-dimensional reconstruction algorithm is likely to be individually inadequate. Some, for example, may be limited by fundamental constraints; e. g., the problem may be ill-posed, with no unique solution possible because sufficient information for surface or range reconstruction is simply not available in the two-dimensional image. That is, when a three-dimensional object is reduced to a two-dimensional image, critical and necessary information or crucial constraints may not be retained. Adding supplementary constraints or "regularizing" the situation may sometimes overcome these procedures (Grimson, 1981; Poggio and Girosi, 1990). On the other hand, some algorithms may be impractical simply because they result in a computational explosion that inevitably occurs when they are pushed to their respective limits. Others may be limited because the sampling of the image must be constrained to an unacceptably low density so that the problem may be computationally tractable. What remains is a sampled surface that is inadequate to represent the surface fully.

Combination approaches overcome these limitations in a manner that is analogous to the use of simultaneous equations in solving algebraic problems. When one combines, one adds constraints on the solutions that may be obtained and often allow one to solve a problem that cannot be solved with only one algorithm. On the other hand, combination approaches such as the one we propose here are subject to one major disadvantage: they are always going to be more computationally intensive than are the individual operators of which they are composed. However, given (1) that individual procedures are not likely to be generally competent and (2) that combination is the method par excellence that seems to have evolved in the organic nervous system, the combination approach taken here seems increasingly appropriate.

In the following sections of this chapter we describe some improvements that we have made in three conventional algorithms that individually convert two-dimensional images to three-dimensional representations. Then we describe a pair of novel combination processes that has been used to unite the three into a more universally competent three-dimensional shape and range recovery system.

6.2 THE COMPONENT ALGORITHMS

6.2.1 Stereo (Two-Camera) Reconstruction Algorithms

Our system simulates the stereoscopic process that occurs in binocular human vision by recording two images of the same solid object from two horizontally separated camera positions. The stereoscopic technique has long been of interest to computer vision specialists and is supported by one of the most highly developed computational modeling traditions. Marr and Nishihara (1978), Marr and Poggio (1979), Marr and Hildredth (1980), and Grimson (1981) all made important theoretical contributions to understanding stereopsis in organisms with their computational models.

There are, however, fundamental constraints at work in any practical computational stereoscopic system prohibiting the ultrafine resolution of depth differences that is characteristic of the human visual system. These limits mean that contemporary computer technology is unlikely to provide precise estimates of the detailed shape of a surface using stereo disparities alone. The problem arises because of a practical trade-off between the resolution of the stored image, the amount of computer time required to execute certain parts of the process, and the size of the computer memory required to store intermediate images.

It was shown by Kakarala that the maximum ability to discriminate between depths (z resolution) is a fixed proportion of the x-y resolution. For practical resolutions (for example, 256×256 pixels or 512×512 pixels) that do not extend the processing time beyond acceptable limits, the best obtainable depth resolutions are only about 2% for a 256×256-pixel image and about 1% for a 512×512 one. These limits assume a total depth range of 500 cm, a viewing angle of 30°, and a sensor width of 1.27 cm. These rather poor depth resolutions, equivalent to disparities of 18 and 36 arcmin, respectively, compare extremely unfavorably with the limits of human depth perception—a value that is usually assumed to be 2 arcsec. For this reason, stereoscopic algorithms are best for determining the range to an object rather than local surface shape. However, as demonstrated later, the stereoscopic information may be otherwise useful.

Stereo analysis is fundamentally a three-step procedure. The first step, the most difficult one, is to solve the *correspondence* problem; that is, to determine which parts of the two two-dimensional images from the pair of cameras are equivalent. The second step is to determine the horizontal shift, or *disparity* (if the two cameras are, like the human eyes, displaced from each other horizontally), of the parts of the image that have been determined to correspond. The third is to use that disparity to compute the relative depth of different parts of the image. This third step is essentially a trivial piece of computational trigonometry. The solution to the initial correspondence problem, on the other hand, is the source of most of the theoretical difficulties and the time-consuming calculations in the computational model.

The algorithm we developed for stereo image processing is based on the idea of "zero-crossing" analysis originally proposed by Marr and his colleagues. The correspondence problem was solved by them in an ingenious manner: the features that are used as interest operators are the points on a gray-scale image where the Laplacian of the Gaussian function of the image equals zero. The locations of these "zero-crossing" points are particularly useful interest features because they are signed operators. That is, a crossing from a bright to a dark region is distinguished from one that crosses from dark to bright. This additional information helps to determine the corresponding contours produced by what is essentially an edge enhancing process.

Our adaptation of this correspondence-detecting algorithm, like theirs, utilizes a procedure in which matching of interest features is carried out at four different levels of resolution. The four progressively finer resolutions produce four different correspondence maps. The findings at each level are then used to help resolve any ambiguities that may still exist at the higher levels. The end product of this iterative process is a final map showing the best estimates of the correspondences of the many zero crossings.

There are, however, some practical computational problems associated with the Marr–Nishihara–Poggio–Hildreth–Grimson approach that led us (specifically Kakarala) to invent some shortcut procedures. For example, rather than calculate the Gaussian transform of the image directly (a process initially required to smooth the original image), a much simpler and less computationally demanding local averaging process was substituted. Furthermore, rather than attempting to calculate the Laplacian directly, we used a difference-of-Gaussians approximation. Although there are some advantages to such approximations (e.g., computational speed in this case is dependent only on the size of the image and, therefore, the process can be conducted in the spatial domain quite rapidly), these shortcuts do not reproduce exactly the specific procedure inherent in the formal mathematics described by earlier workers. Nevertheless, they do work sufficiently well for the purposes of our model.

It is important to emphasize that these approximations are not the source of the limits on depth resolution of a stereo system described previously. Rather, those limits are properties of the pixel count and processing speed of the computer and, therefore, are fundamental rather than methodological. There is no algorithm that could overcome the depth resolution limit, given a particular pixel count.

Since the other two algorithms—SFS and SFSL—are excellent at measuring the details of surface shape but poor at defining the range, the combination of their respective capabilities to recover surface detail with the stereo algorithm is particularly advantageous.

Finally, the values of depth obtained from the Stereo algorithm are sparse and occur only where the zero crossing points are located. Figure 6.1 is a needle diagram of a typical output of the Stereo system. The depth of a particular region is determined by averaging all of the points in that region. This assumes that the surface is planar and fronto-parallel, an assumption that may not be correct but that introduces only minor errors in a range estimate. In addition, since stereo range estimates are susceptible to error due to noise or the intrinsic uncertainties due to miscorrespondences of the images from the two cameras, minor discrepancies in ranging due to surface irregularities are usually inconsequential.

6.2.2 Shape from Structured Light (SFSL)

Our first approach to determine the fine details of the local shape uses projected patterns of two kinds: a pattern of horizontal, parallel, straight stripes and a similar pattern of vertical, parallel, straight stripes. Our approach is a modification of the method initially suggested by Wang et al. (1987). The SFSL operator depends on the distortion of the projected stripe pattern by the geometry of the surface of the object. A sample of the distortions of the projected vertical and horizontal stripes produced by a hemispherical object is shown in Fig. 6.2. The computational task is to compare the distorted pattern with an undistorted pattern and to use the differences between the two as a cue to the surface shape of the object. The process, therefore, depends on the identification of equivalent points on the distorted and undistorted stripe patterns.

Our modification of the Wang et al. (1987) procedure involved two funda-

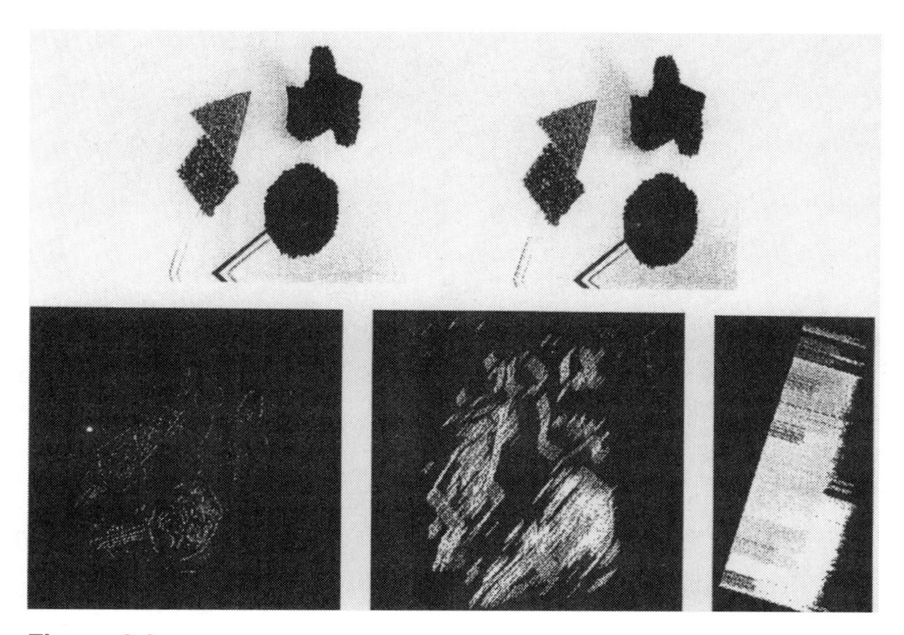

Figure 6.1 This figure shows the input and output from the Stereo algorithm described in this chapter. The top two photographs are the left and right camera views, respectively, of a simple scene consisting of four colored, textured surfaces at different distances from the cameras. (Cross-eyed fusion reveals the circular object to be closer than the other three.) The three lower pictures are views (top, oblique, and side) of different orientations of the needle diagram resulting from the execution of the algorithm. The tips of the needles represent estimates of the range of a point x,y from the background (zero disparity) surface. These pictures illustrate two inadequacies of the Stereo algorithm. First, the three lower pictures demonstrate the relatively sparse depth estimates obtained. Second, the side view demonstrates the randomness of the obtained depth estimates, even within the confines of the patch of needles standing out from the others.

mental changes. First, in their original paper these authors required that the object be placed on a physical base plane to provide a reference for the slope and separation values of the undistorted stripes. We simplified this process by computing the stripe pattern that would have fallen on a virtual, but nonexistent, plane at the average depth of the surface, rather than actually having a physical base plane present. This was accomplished by computing the global average of the virtual slopes and the average virtual stripe separation from the actual distorted stripes reflected from the object—a process that is described in greater detail later. Since the reference, undistorted virtual stripe pattern is computed

Figure 6.2 A sample of the output of the SFSL algorithm. The left-most picture is a photograph of the original object—a basketball. The two faint sets of lines to its immediate right show the actual image picked up by the camera for the horizontal and vertical projected stripes. The next pictures to the right show the interpolated and visualized reconstructions based on the vertical and horizontal stripes, respectively.

from the actual image, the object can be free floating in space. This is an important advantage over preceding work.

The second improvement was a set of preprocessing steps that improved the quality of the images of the parallel stripes. Because of depth-of-focus limitations from any practical projector, surface discontinuities on the object, or the discrete processing algorithms themselves, the stripes often are discontinuous, vary in width, or are otherwise imperfect representations of the fine stripes originally projected. Some of our preprocessing steps used appropriate thresholding and skeletonizing procedures to produce a one-pixel-wide stripe, but one that often was broken into incomplete segments.

The method we have developed to overcome this difficulty is obviously not the only way in which this could be done. A simple least-squares fit might work in some instances. Nevertheless, our method would be expected to be more robust for larger breaks in the lines. It must be acknowledged, however, that no technique of this sort could possibly work all of the time for all kinds or magnitudes of line interruptions.

To link broken stripe segments in our procedure, each segment was individually identified with a unique code. Segments were clustered on the basis of a recursive algorithm that grouped adjacent pixels into the same stripe segment. Isolated pixels and stripes shorter than an arbitrary threshold (usually equal to 10 pixels) were then removed.

The closing of broken stripe segments is not without its difficulties, because of the distortions produced by the object surface. Figures 6.3 and 6.4 show two situations in which continuity is ambiguous. The following two algorithms have been successfully applied to resolve these kinds of ambiguities and to connect the ends of the stripes properly.

Connection Algorithm 1

The endpoints of each stripe in the image are identified, and a list of all the endpoint coordinates is constructed. Then, for each stripe S_i, the upper endpoint coordinate value is compared with the lower endpoint coordinate values of all of the other stripes. If the lower endpoint of another stripe is within a threshold range of the upper endpoint of the stripe being tested, then the latter stripe's ID is reset to the former stripe's ID. This process is repeated for each stripe (S_i) until all are tested. This process is carried out separately for horizontal and vertical stripes.

Figure 6.3A shows the case for horizontal stripes. Here, since the endpoints of stripes S_1 and S_4, namely, (x_2,y_2) and (x_3,y_3), are within the bounding box

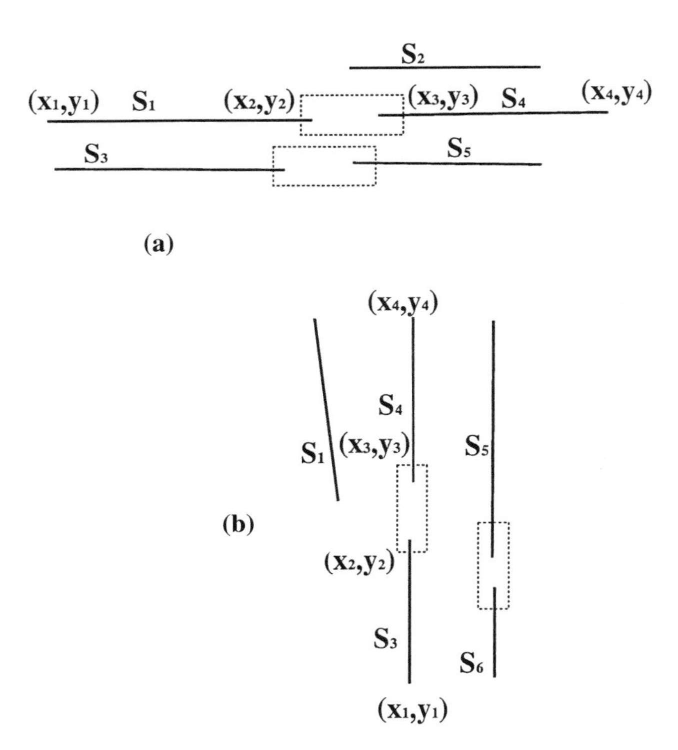

Figure 6.3 Diagram illustrating how the first line-closing algorithm works for horizontal (a) and vertical (b) straight lines, respectively. Captions and process description are in the text.

representing the threshold limits (in both the X and Y directions), the two stripes are treated as parts of a single stripe. Figure 6.3B shows the case for vertical stripes.

Connection Algorithm 2

Not all gaps between stripe segments are closed by the first procedure. Therefore, the image is next passed through a more elaborate algorithm for joining any remaining broken stripes. In this second algorithm, a list of all the endpoints of the remaining unconnected stripes is constructed. Then, for each stripe, S_i, the slope m, and the midpoint $M(X,Y)$ are calculated using its endpoints. Figure 6.4A shows this process for a horizontal stripe image. If x_3 (the x-coordinate of the lower endpoint of stripe S_2) is greater than x_m (the x_2-coordinate of the midpoint of stripe S_1), then a point $P(X,Y)$ is found such that it lies in the path of the stripe S_1 with the coordinates calculated as follows:

$$M(x) = \frac{x_1 + x_2}{2}$$

$$m = \frac{y_2 - y_1}{x_2 - x_1}$$

$$P(X) = x_3$$

$$P(Y) = m(x_3 - x_1) + y_1$$

If the range between $P(Y)$ and y_3 is less than a threshold limit T, then S_1 and S_2 are given the same ID number. After those two stripes are joined, a new midpoint and a slope are computed using the outermost endpoints of the joined stripe. The procedure is repeated until no more stripes can be joined.

Similarly, in Fig. 6.4B, the vertical stripe image, the following equations are used to find the coordinates for point P, and the corresponding stripes are joined if the range between $P(X)$ and $S_2(x_3)$ is less than the threshold limit T:

$$M(y) = \frac{y_1 + y_2}{2}$$

$$m = \frac{y_2 - y_1}{x_2 - x_1}$$

$$P(X) = \left(\frac{y_3 - y_1}{m}\right) + x_1$$

$$P(Y) = y_3$$

The reconstructed stripes are then sorted according to their relative ascending

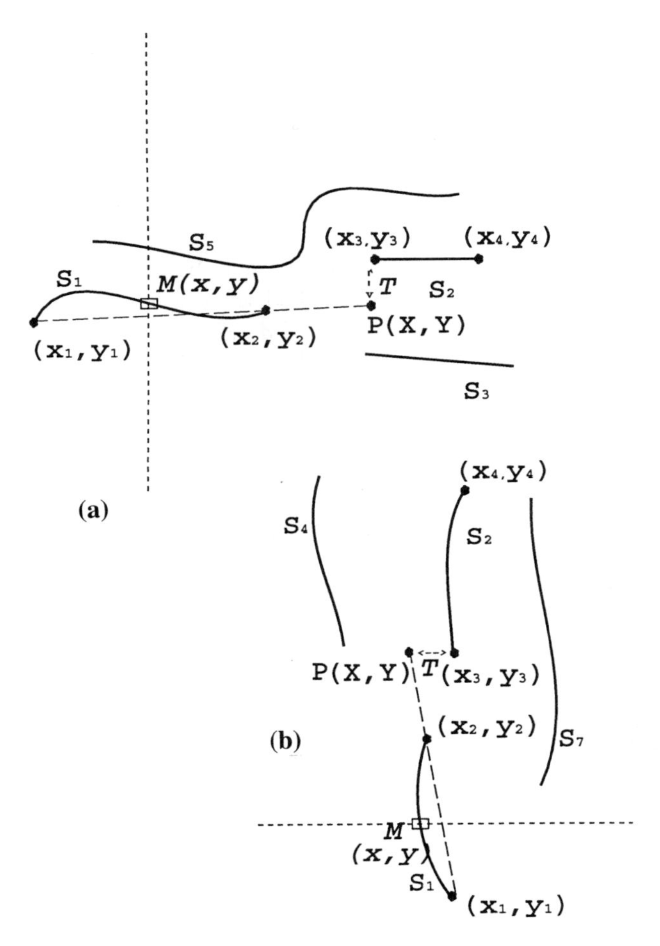

Figure 6.4 Diagram illustrating how the second line-closing algorithm works on curved horizontal (a) and vertical (b) lines, respectively. Captions and process description are in the text.

order for horizontal stripes and from left to right for the vertical stripes. The sorting uses the coordinate values of the endpoints as the criterion.

Surface depths are then computed using the geometric deformation of the projected grid (originally a regular pattern of parallel stripes) as a cue. If a nonplanar object is in place instead of a planar surface, the stripes are distorted by the interaction between the geometry of the surface and the set of equally spaced projected stripes. This pattern of distorted stripes is designated as the actual stripe image since they are "actually" imaged by the camera. The coordinate values of all the points along the set of virtual stripes are then computed. The distances between corresponding points along the actual stripes and the computed points along the virtual stripes are used to calculate the depth at each point on the virtual stripes.

Figure 6.5 shows an image of a horizontal cylinder with a regular pattern of vertical stripes projected onto it. Using this figure, the procedure can now be explained more fully.

First, the slope and separation of the virtual stripes must be computed by calculating the average distances for each point on all adjacent distorted stripes in the actual image. These average distances are coded $d_1, d_2, \ldots, d_n$, where n is the total number of stripes in the image. A global average range D is then calculated as follows:

$$D = \frac{d_1 + d_2 + \cdots + d_n}{n}$$

Second, a global average slope M is calculated by taking the average of the

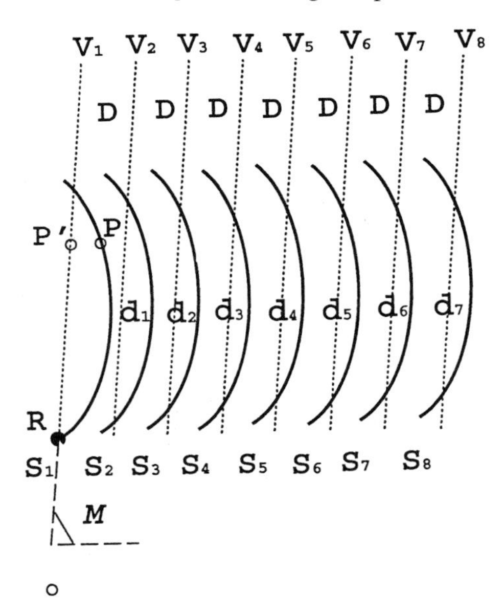

Figure 6.5 Diagram illustrating how the SFSL computations reconstruct a cylinder. S_n represents the ID number of each real line observed in the shape-distorted image. V_n represents the ID number of the virtual stripes with common average slope M and separation and average range d_i between points on stripes S_i and S_{i+1}. D is the grand average range of all the d_i virtual stripes. R is the reference point relative to which depth is computed. P' and P are corresponding points on the virtual and real stripes, respectively.

individual slopes of all distorted actual stripes using their respective endpoints as the criteria. Virtual stripes, $V_1, V_2, \ldots, V_n$, corresponding to the distorted stripes $S_1, S_2, \ldots, S_n$ that are actually imaged, can now be generated with a uniform slope M. The process starts from a reference point R equal to the lowest point of the stripe having the lowest ID number. The range between adjacent virtual stripes is then defined to be exactly D.

The depth of each point x,y ($\text{Depth}_{x,y}$) is then computed by taking the range Δd between a point P_n on the actual stripe and the corresponding point P'_n on the virtual stripe. This range is multiplied by an appropriate scaling factor γ. A depth map relative to the virtual planar background is then constructed by adding the scaled range to the virtual background, as represented by:

$$\text{Depth}_{x,y} = \text{SURFACE\_THRESHOLD} + \gamma\,\Delta d$$

SURFACE_THRESHOLD is the level of the virtual reference plane. Initially, depth values are set to zero. After the evaluation of this expression, depth values for those points on the object have a base value equal to the value of SURFACE_THRESHOLD. All other points on the background retain their zero depth values. Because SURFACE_THRESHOLD does not affect the actual shape of the object, one can be arbitrarily picked to provide the best visualization effect. The scale factor γ depends on the optical arrangement of the slide projector and camera. Once their relative position is fixed, γ is a constant.

One other approximation is used in the algorithm. Each acquired stripe image is a very dense line of pixels. Theoretically, depth values could be computed at every pixel on the stripe. Since the stripes themselves are so sparse (adjacent stripes are separated by 10–25 pixels), processing all of this information would be computationally wasteful. Therefore, our algorithm examines only every tenth pixel on each stripe to produce the SFSL depth estimates. These relatively sparse sets of depth estimates are then interpolated in both the x- and y-directions and a surface fitted for visualization using our surface rendering system. A similar procedure is used for horizontal stripe images.

Though efficient and very successful in many cases (as shown in Figs. 6.2 and 6.6) in reconstructing the surface of an object, this technique for computing shape from projected structured light has a number of disadvantages. First, it is intrinsically a sampled procedure. That is the stripes are not dense enough to reconstruct the entire surface. Surface interpolation methods (as illustrated in Figs. 6.2 and 6.6) are necessary to fill in the voids and reconstruct the surface. Second, although able to distinguish local depth differences and thus extract shapes, it is not capable of determining the absolute range to the entire object. Third, it is an active process; that is, a grating must be projected onto the objects. Although this final disadvantage is not critical to our model and would not be a handicap for a self-textured surface, it is a practical consideration.

Figure 6.6 A sample of the results of the SFSL algorithm for a piece of acoustic foam. The sequence of pictures is the same as in Fig. 6.2 except that two additional views of the reconstructed surface are presented.

6.2.3 Shape from Shading (SFS)

The purpose of the SFS algorithm is the same as the SFSL procedure just described: to reconstruct the detailed three-dimensional surface shape of an object from a two-dimensional image. The necessary cue in the case of SFS, however, is much more subtle and is dependent on factors other than just the shape of the surface. Shading is a function of not only the surface shape of the object but also the illuminant direction, its ray pattern, the reflectance function, and the albedo of the surface. A further complication is that a two-dimensional shaded picture does not unambiguously contain all of the information necessary to determine if the object is concave or convex. Therefore, by itself the SFS algorithm has multiple satisfying solutions. It is, therefore, said to be ill-posed.

Due to the fact that the general shape from shading problem is ill-posed mathematically and that the mathematics are relatively difficult to solve, several constraining assumptions are usually made in constructing a practical mathematical model and an algorithm. As Horn and Brooks (1989) and Blake et al. (1989) point out, if one assumes that the light source is distant and that light rays are nearly parallel, then the brightness of a region on a surface depends only on its orientation and not on its range to the light source. Therefore, a simple orthographic projection (parallel rays) lighting model, rather than a more complex perspective projection (diverging rays) lighting model, can be used.

There has been considerable recent activity in exploring how shape can be reconstructed from shading by iterative methods. In general, iterative methods

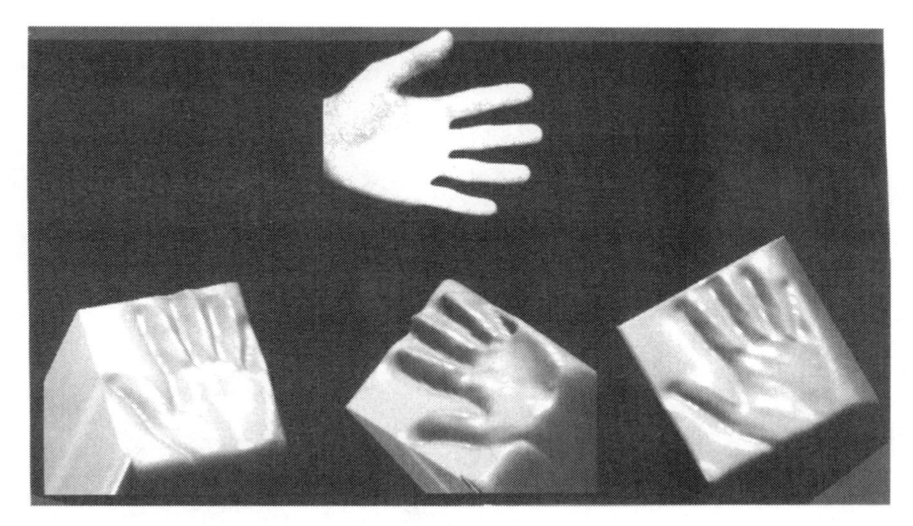

Figure 6.7 A sample of the SFS algorithm for a hand. The top picture is the original two-dimensional image captured by the computer. The bottom three pictures are different views of the three-dimensional surface reconstructed by the algorithm.

repeatedly make adjustments in some ad hoc surface shape until the computed shading matches that actually observed in the image. Horn (1970, 1990) is usually credited with the development of the earliest computational methods for carrying out the transformation. Pentland (1984) provided an improved method for determining the illuminant direction, and a further improved version was reported by Lee and Rosenfeld (1989). The algorithm we use here is based mainly on a recently developed iterative method by Zheng and Chellappa (1991). As these authors summarize, a practical solution for the SFS problem requires the execution of a number of steps.

First, the lighting direction (i.e., the slant and tilt angles of the illuminant) relative to the object coordinate system must be computed. In this regard, we must point out again that all formal models of this process depend on strict characterization of the lighting. For example, the current model is based on an illuminant that is composed of parallel rays, i.e., a point source at an infinite distance. In passing, however, we note that the modified algorithm we wrote is much more robust than this strict criterion would have suggested. For example, diffuse lighting—ordinary overhead fluorescent lighting—does provide an adequate illuminant, as demonstrated in Figs. 6.7 and 6.8.

The second step is the determination of the albedo of the surface. This part of the calculation also depends on a fairly strict assumption—a Lambertian surface. That is, the light reflected from any point is a cosine function of the angle

Figure 6.8 A sample of the SFS algorithm for another type of acoustic foam, but processed by the SFSL. The upper left picture is a photograph of the material. The upper right-hand picture shows the original acquired two-dimensional image. The lower two images are three-dimensional reconstructions of the surface from the acquired image.

between the normal to the surface and the viewing angle. Again, using a variety of materials of varying kinds of reflectance, we found the algorithm to be more robust than the rigorous model suggested. This finding had been anticipated by Woodham (1980), who showed that the Lambertian model was also surprisingly effective for aerial photographs, even when the reflection pattern did not actually meet this strict criterion.

The third step is the actual computation of the reflectance map of the surface using the assumptions of a Lambertian model, a parallel, remote light source, and a uniform albedo, albeit with more complex input images than those assumed in the model. The reflectance map is a measure of the amount of light coming from each point on the surface from the point of view of the observer or camera. Given this information, it is possible to determine the surface normal at each point on the observed surface.

This brings us to the fourth step, described by Zheng and Chellappa (1991). To reconstruct the surface, the surface normal and the depth value at each point must be determined by a hierarchical or pyramidal iteration scheme; that is, like the algorithm designed to determine stereoscopic image correspondence, the reconstruction is best carried out at several levels of resolution. The levels of resolution used in our version of the algorithm are 64×64, 128×128, and 256×256 pixels. At each level, the reconstruction is accomplished by an iterative procedure until convergence is achieved or an upper limit on the number of iterations is exceeded.

The purpose of the multilevel pyramid is efficiency. Approximate, though rough, depth solutions can quickly be obtained for a few points on a low-resolution version of the image. These solutions can then be used as starting points (seeds) for the next-higher-resolution level. With an appropriate seed as an initial condition, the iterative procedure converges much faster than when starting with an initial flat field of zero depth values. Shape information is communicated upward from one resolution level to the next higher one by interpolating between existing values to ones that would have been empty in the higher level.

Although we followed their published procedure in general, the following modification was made in the Zheng and Chellappa (1991) algorithm, an improvement that they suggested but did not implement. We used a cost function with values of $\lambda < 1$, rather than setting it to 1 as they had done:

$$\iint F(p,q,z)\, dx\, dy$$

where

$$\begin{aligned} F = {} & [R(p,q) - I(x,y)]^2 \\ & + \lambda[R_p(p,q)P_x + R_q(p,q)Q_x - I_x(x,y)]^2 \\ & + \lambda[R_p(p,q)P_y + R_q(p,q)Q_y - I_y(x,y)]^2 \\ & + \mu[(p - Z_x)^2 + (q - Z_y)^2] \end{aligned}$$

The empirical formula for the progressively decreasing weighting constant λ we used was

$$\lambda_i = e^{-i/20}$$

where i is the number of the current iteration. Otherwise, our version of this algorithm was consistent with that described by Zheng and Chellappa (1991, p. 668).

6.3 THE COMBINATION PROCEDURE

So far we have described three different and separate algorithms for determining the distance, depth, or surface shape of an object. Each of these algorithms is capable of producing some information about the *z*, or third, dimension of a scene. In some cases, the reconstruction of surface shape or range can be very good. However, each is limited in some respects. No one of these algorithms is capable of completely and robustly producing a universally competent three-dimensional reconstruction of a wide variety of two-dimensional images under all conditions and in all situations. It has almost become an axiom of computer vision that only rarely does an algorithm generalize outside of the specific microuniverse of images for which it was originally designed. Small deviations in image type may require completely different assumptions and programs to be processed.

In exactly the same way and as we extensively discussed in Chapter 2, it is now clear that no single cue can provide a robust and general visual perception of a three-dimensional scene by an organism. Unconstrained operators, whether they be organic or in a computer, produce distorted, unstable, or ambiguous responses. In human visual perception these nonveridical distortions are called *illusions*. The general conjecture that there must be heavy integration and interaction between different attributes of a visual scene can be particularized to the visual corollary that depth and surface shape also require multiple cues to produce a compelling and unambiguous perception of the third dimension of space. The model we now propose is based on this conjecture. It seeks to simulate some of the functional process capabilities of the mechanisms that underlie visual perception, not the specific neural mechanisms themselves. Our specific goals are to reconstruct a unique, robust, and accurate estimate of the surface shape of an object and to determine the absolute range from the object to the camera.

6.3.1 Some Assumptions

We have argued that the goal of determining range and surface shape cannot be accomplished in a dependable way using in isolation any one of the three algorithms described so far. Each of the three methods (Stereo, SFSL, and SFS) depends on only a single cue or attribute of the visual scene. Each is beset by certain limitations and constraints that make its determinations of the third dimension either questionable, ambiguous, or incomplete. Aloimonos and Shulman (1989) (p. 20) suggest three general problems with these three-dimensional reconstruction algorithms:

the nature of the basic computational assumptions
the uniqueness of the computational solutions
the robustness or stability of the algorithms

Aloimonos and Shulman's first problem is well illustrated by the following surprising result. We have found that the SFS algorithm does not work with very simple objects such as a hemisphere. This is due entirely to the nature of the mathematical assumptions built into the Zheng and Chellappa model. Their algorithm fails for a hemisphere because their model has been linearized; they expand the reflectance map up to linear terms only. A hemisphere produces significant nonzero high-order terms because of the very steep slope at its edges. Hemispheres and other objects of similar simple shape are examples of the kinds of objects that tend to be falsely reconstructed as flat planes by the SFS algorithm. (It should also be noted in passing that this difficulty can be overcome only at enormous cost in computing time.)

Consider next the following example of a nonunique solution. Because the SFS algorithm is based on an iterative process whose convergence depends on the minimization of a function, a unique solution does not always obtain. In other words, the algorithm may be ambiguous or a solution may converge to a false minimum. The reason for this ambiguity is that the SFS process is trying to solve an ill-posed mathematical problem; the original two-dimensional image simply does not contain all of the information necessary for the reconstruction of the surface shape. Typically, two satisfactory solutions can be obtained from the SFS algorithm. One potential solution corresponds to a convex surface, although the other corresponds to a concave surface. Both represent possible convergent solutions satisfying the problem posed by the two-dimensional image. Looking at the problem from the opposite direction, either a concave or a convex three-dimensional surface could produce the same shaded two-dimensional image. Ipso facto, the information remaining in the two-dimensional image is necessarily insufficient to reconstruct the three-dimensional surface without the addition of supplementary constraints and assumptions.

It is important to appreciate that this algorithmic limitation is also characteristic of human vision. Faced with a two-dimensional picture that provides ambiguous shading information, the human observer can perceive the figure as either convex or concave, depending on the additional assumption of the direction from which the light is coming. This information is not implicit in the image and is completely independent of the physical properties of the image. Yet this high-level "cognitive" information is perfectly capable of reversing the perception of a scene. A compelling demonstration of this phenomenon is shown in Fig. 6.9 (from Ramachandran, 1988) In the case of human vision, the additional assumptions or constraints are applied to the raw image data by central neural processes of which we know very little. In the case of the SFS computational algorithm, they must be applied by the programmer.

Finally, computational algorithms of this kind are sometimes extremely fragile. For example, in the SFSL algorithm, if only a single stripe in the reflected image returning from an object is missing, a huge distortion in the reconstructed

Figure 6.9 A demonstration of the effect of assumed lighting direction in human perception. Humans, like computer programs, operate under constraining assumptions. Here the lighting is assumed to come from above. If the picture is turned upside-down, what been perceptually convex becomes concave and vice versa. (After Ramachandran, 1988).

three-dimensional shape is produced. It is for this reason that we have paid such extreme attention to the completion of broken line segments and the production of a "clean" virtual image.

Similarly, the distribution of textural elements on an object being examined by the stereoscopic camera procedure can significantly alter the measurements produced by that algorithm. If, for example, the component features of the object are even approximately periodic, false correspondences are produced that generate grossly distorted range measurements. It is, for this reason, that randomly textured surfaces are best used for this process.

There are also difficulties with these algorithms associated with noise in the image. This is particularly characteristic of the Stereo process. This algorithm produces a "needle map" of depth values (see Fig. 6.1), many of which are the results of spurious identifications of what are incorrectly thought to be corresponding points. Even properly localized needles within a plane can vary substantially in their height, due to vagaries in the calculation process, as is also illustrated in Fig. 6.1. This intrinsic variation makes precise depth measurements irregular at best. This means that this process is inappropriate for determining the detailed surface shape of an object at any great distance. The Stereo algorithm, therefore, in our model is used only to determine the rough range to an object. It is important to remember that this is not a methodological flaw in the Stereo algorithm, but rather an intrinsic problem arising from the practical pixel densities that can be used in the procedure.

Quite to the contrary, both SFS and SFSL are inadequate in determining absolute distances of an object from the camera. However, since they are effective at determining the slight depth differences that are characteristic of surface shape, they can be used for precise shape evaluations and thus complement the strengths and weakness of the Stereo algorithm.

Finally, it must also be appreciated that these three algorithms for determining depth and range sometimes produce solutions that are incomplete in one way or another. That is, only a relatively few points may be sampled on the surface. Shape from shading produces the most complete representation of the surface (depending on the degree to which one wishes to load up one's computer), but both stereopsis and shading from structured light produce their solutions as a set of sparse samples. Interpolation procedures are necessary for all of these methods, to a greater or lesser degree, to produce a smooth visualization or perception of the surface. The samples shown, for example, in Figs. 6.2, 6.6, 6.7, and 6.8 have been completed (i.e., interpolated) using a visualization system developed in our laboratory by Dayanand.

The incomplete or sampled solutions produced by the stereoscopic and SFSL algorithms do permit them to be computationally quite efficient compared to a method that might attempt to use all of the pixels in the image. Thus, either or both may serve the same function as a seed in the SFS algorithm as did one of the lower-resolution images developed by the SFS algorithm itself.

The situation engendered by the fact that each of these algorithms has its own inherent constraints and limitations can be at least partially handled by combining them in various ways. We have argued throughout this book that it should be possible to produce an integrated output of several algorithms that is superior, more valid, and more robust than that produced by any one alone. In particular, we now describe how SFS and SFSL are combined to give a robust surface reconstruction and SFSL and Stereo are combined to produce robust range information.

6.3.2 Combining SFS and SFSL

The difficulty with the SFS algorithm, as indicated earlier, was the ill-posed or ambiguous nature of the two-dimensional image. A shaded image simply does not contain sufficient information to resolve the difference between convex and concave objects. The iterative solution can properly converge on equally plausible alternative solutions. Similarly, the SFSL algorithm is limited by practical limits on the density of the data that can be processed. However, SFSL is not ambiguous; it is well-posed mathematically. Therefore, our approach has been to combine SFS and SFSL by using the SFSL to resolve the ambiguous output of the SFS algorithm. Our decision to proceed in this direction was also based on the following additional considerations:

Estimated depth values in the SFS algorithm can also be exaggerated or mitigated as a result of variations in image smoothness or albedo.

SFS is perturbed by backgrounds. The entire process depends on uniformity of the lighting pattern, reflectance functions, and reflectance parameters. Backgrounds with different textures can seriously affect the outcome.

SFSL is extremely sensitive to missing points on the projected lines or to entire missing lines.

The structured light used in SFSL can be constrained to the object itself.

The algorithm we developed to combine SFS and SFSL consisted of the following steps:

Step 1: The boundary of the object image is determined from the SFSL algorithm.

Step 2: The depth values from the SFS algorithm outside the SFSL boundary are declared to be “invalid”—i. e., they are not considered in the computation.

Step 3: The result of SFSL algorithm is normalized by taking the average of the deviations of all computed depth values from their average value. The same is done for all valid depth values obtained from the SFS algorithm. A scale factor is then determined from the ratio of these two normalized average values. This permits the results from the two algorithms to be combined at the same scale.

Step 4: Potentially ambiguous depth values inside the SFS are set to reasonable normalized values using the SFSL values and the scale factor.
Step 5: All "invalid" depth values (outside the SFSL boundary) in the SFS are set to the lowest scaled depth value.

These five steps are now described in greater detail in the form of a pseudoalgorithm. Specifically, the result of the SFSL algorithm is in the form of a set of triples representing the surface:

$$R_{\text{sfsl}} = (X_i,\ Y_j,\ Z_{ij})$$

where $i = 0, \ldots, M, j = 0, \ldots,$ N, $X_O \le X_i \le X_{M'}$ and Z_{ij} is the depth value at position (X_i,Y_j).

The result of shape from shading is a similar set of triples, of the following form:

$$R_{\text{sfs}} = \{(\tilde{X}_1,\ \tilde{Y}_1,\ \tilde{Z}_1)_{1=0\ldots L}\}$$

where $\tilde{Z}_1$ is the depth value at position $(\tilde{X}_1,\tilde{Y}_1)$.

Step 1: Boundary Determination. R_{sfsl} is used to find the boundary of the object:

$$Y_{min\_i} = \min(Y_j) \text{ where } (X_i,Y_j,Z_{ij}) \in R_{sfsl}$$
$$Y_{max\_i} = \max(Y_j) \text{ where } (X_i,Y_j,Z_{ij}) \in R_{sfsl}$$

Step 2: Establishing the Validity of the SFS Points. The following set of rules is used to determine whether a point (X,Y) is inside the boundary:

1. If $X < X_0$ or $X > X_M$, then (X,Y) is outside the boundary.
2. If $X = X_i$ and $(Y < Y_{\text{min}\_i}$ or $Y > Y_{\text{max}\_i})$ for i=0 . . . M, then (X,Y) is outside the boundary.
3. If $X = X_i$ and $Y_{\text{min}\_i} \le Y \le Y_{\text{max}\_i}$, then (X,Y) is inside the boundary.
4. If $X_{i-1} \le X \le X_i$ and

$$Y < Y_{\text{min}\_i-1} + (Y_{\text{min}\_i} - Y_{\text{min}\_i-1}) \times \frac{X - X_{i-1}}{X_i - X_{i-1}}$$

or

$$Y > Y_{\text{max}\_i-1} + (Y_{\text{max}\_i} - Y_{\text{max}\_i-1}) \times \frac{X - X_{i-1}}{X_i - X_{i-1}}$$

then (X,Y) is outside the boundary.
5. If $X_{i-1} \le X \le X_i$ and

$$Y_{\min\_i-1} + (Y_{\min\_i} - Y_{\min\_i-1}) \times \frac{X - X_{i-1}}{X_i - X_{i-1}}$$
$$\leq Y \leq Y_{\max\_i-1} + (Y_{\max\_i} - Y_{\max\_i-1}) \times \frac{X - X_{i-1}}{X_i - X_{i-1}}$$

then (X,Y) is inside the boundary.

Applying this set of rules to all $(\tilde{X}_1,\tilde{Y}_1,\tilde{Z}_1) \in R_{\text{sfs}}$:

If $(\tilde{X}_1,\tilde{Y}_1)$ is inside the boundary, then $\tilde{Z}_1$ doesn't change.
If $(\tilde{X}_1,\tilde{Y}_1)$ is outside the boundary, then $\tilde{Z}_1$ is reset to INVALID.

Step 3: Normalization and Scaling. To scale the inner depth values obtained from the SFS algorithm using the result of the SFSL algorithm, the average deviational value of each set is determined as follows:

$$Z_{\text{average}} = \frac{\sum_{i=0\ldots M,\, j=0\ldots N} Z_{ij}}{N\{Z_{ij}\}}$$
$$\|Z\|_1 = \frac{\sum_{i=0\ldots M,\, j=0\ldots N} |Z_{ij} - Z_{\text{average}}|}{N\{Z_{ij}\}}$$

where $N\{Z_{ij}\}$ is the number of points inside the boundary, Z_{average} is the raw average depth value, and $\|Z\|_1$ is the average deviation from that raw average—the average deviation from the raw average depth value. The latter computation is required because the base level of the SFS and the SFSL may differ depending on the physical and initial value settings of the instrumentation.

$$\tilde{Z}_{\text{average}} = \frac{\sum_{l=\{k|k=0\ldots K,(\tilde{X}_k,\tilde{Y}_k)\}} \tilde{Z}_l}{N\{\tilde{Z}_l\}}$$
$$\|\tilde{Z}\|_1 = \frac{\sum_{l=(k|k=0\ldots K,(\tilde{X}_k,\tilde{Y}_k)\}} |\tilde{Z}_l - \tilde{Z}_{\text{average}}|}{N\{\tilde{Z}_l\}}$$

for all $\tilde{X}_k,\tilde{Y}_k$ inside the boundary. Here, $\tilde{Z}_{\text{average}}$ and $\|\tilde{Z}\|_1$ play the same role as just described, but for the SFS algorithm.

The scaling factor for shape from shading can now be specified as:

$$\text{scale} = \frac{\|\tilde{Z}\|_1}{\|Z\|_1}$$

Step 4: Scaling the SFS Depth Estimates. To compute the final depth values, the following formulae are applied to scale the SFS depth values:

$$\tilde{Z}_{l\_\text{final}} = (\tilde{Z}_l - \tilde{Z}_{\text{average}}) \times \text{scale} + \tilde{Z}_{\text{average}} \qquad \text{if } \tilde{Z}_l \neq \text{INVALID}$$
$$\tilde{Z}_{l\_\text{final}} = (\min_{\tilde{Z} \neq \text{INVALID}}(\tilde{Z}_l) - \tilde{Z}_{\text{average}}) \times \text{scale} + \tilde{Z}_{\text{average}} \qquad \text{if } \tilde{Z}_l = \text{INVALID}$$

Step 5. All invalid points are then set to the lowest value of the set of valid points. This appears to place the object in contact with a flat background when the output is visualized. The final depth values are then used to complete the X,Y,Z triple and the results plotted using Dayanand's visualization technique. This procedure produces a robust estimate of the surface as shown in our example figures. But a range estimate is still needed. This is the purpose of the next section.

6.3.3 Combining SFSL and Stereo

Because of the substantial amount of noise and spurious correspondence matches in the Stereo algorithm, estimated object distances can be quite variable and noisy. Because range estimates have to be derived from a limited area (the segmented object, the major problem is the definition of the object boundaries. Since the SFSL effectively defines the boundaries of objects, it provides a means of overcoming this difficulty. This allows us to take the statistical average of the depths of the sampled points (needles) produced by the Stereo algorithm only within this constrained boundary as a good measure of the range to the object.

The result of the Stereo algorithm is also in the form of a set of noisy triples: the coordinates of the tips of the needles shown in Fig. 1:

$$R_{\text{Stereo}} = \{(\overline{X}_k, \overline{Y}_k, \overline{Z}_k)_{k=0,\ldots,K}\}$$

However, in this case, prior to the disparity calculation $\overline{Z}_k$ is not an absolute depth value. Rather, it is the disparity value computed from the left and right images. If $(\overline{X}_k, \overline{Y}_k)$ is outside the boundary of the object of interest, it may represent a very large miscorrespondence or a spurious point. It may also represent an estimate of the depth to the background or some other object rather than to the object of interest. Since the range desired is from the object to the camera and not from the camera to the background, we declare these points to be invalid and exclude them from the calculation. Therefore, the same set of rules that were used to constrain the SFS depth estimates to the SFSL determined boundaries described earlier can be used in this case:

$$R_{\text{stereo\_valid}} = \{(\overline{X}_k, \overline{Y}_k, \overline{Z}_k) \mid (\overline{X}_k, \overline{Y}_k, \overline{Z}_k) \in R_{\text{Stereo}} \; \textit{for all} \; (\overline{X}_k, \overline{Y}_k) \text{ inside the boundary}\}$$

Only those disparities from the set $R_{\text{stereo\_valid}}$ are now used to compute the range from the object to the camera. This is done by averaging all of the $\overline{Z}_k$ values to give a good estimate of the mean disparity Z'_k. This value is used to produce a good estimate of the range from the camera to the object by applying the following expression for the range d:

$$d = f - \frac{f}{h}\bar{Z}'_k$$

where f is the range from the camera to the point of zero disparity and h is the camera separation. These two constants are determined when the cameras are set up.

6.3.4 Summary of the Combination Algorithms.

Figure 6.10 shows the functional interactions between the SFS and the SFSL algorithms. The combined algorithms can provide a robust estimate of the surface shape of a three-dimensional object extracted from two-dimensional images. Figure 6.11 shows the interactions between the SFSL and Stereo algorithms. The first module in the combined model shown in Fig. 6.12 (a map of the entire combination process) is a UNIX script—"IIScript"—that guides the capture of the required 2D images from the cameras. The acquired images are then transferred to SFS and SFSL preprocessing algorithms. The results from the SFS and SFSL modules are combined in the "combine_and_triangulate" module to produce a robust, unique, and high-resolution surface map as well as to generate a triangulation data file to meet the needs of the surface reconstruction process carried out in the next module—"disptri."

Although the SFS and SFSL combination process was being executed, the images from the two cameras were also sent directly to the Stereo module. The output from the Stereo module was then combined with the SFSL output in the "clean_noise_and_scale" module shown in Fig. 6.12. This module produces a modified disparity file that is used in the module "find_distance" to compute the range from the camera to the object. The last module, "disptri," is a surface reconstruction and display module. It reconstructs a surface from the triangulation data file and allows us to visualize the results of the entire system.

6.4 CONCLUSIONS

The work reported in this chapter is based on the principle that guides all of this book. That is, that an improved computational model of depth perception can be achieved better by combining what are often individually inadequate algorithms than by trying to improve the performance of any individual algorithm. Using this approach we have added to our two-dimensional model a powerful and robust model of a three-dimensional vision system that is capable of excellent range determination and image surface reconstruction. The major disadvantage of this approach, like any other combination approach, is the extensive computational load introduced by the network of interacting algorithms. However, as we have repeatedly pointed out, for a theory of vision this fact is

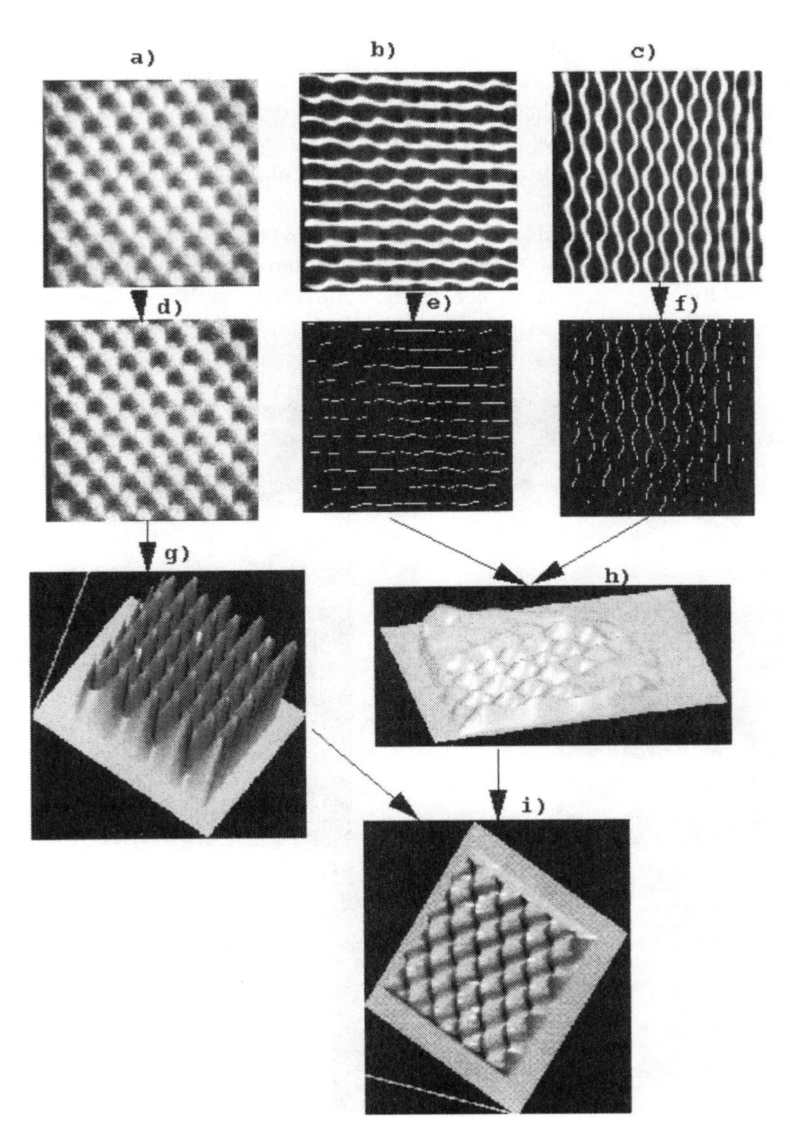

Figure 6.10 A diagram showing the functional interactions between the SFS and SFSL algorithms when combined to reconstruct the surface of a piece of foam rubber. The SFS and the combined (horizontal and vertical) SFSL results are both distorted in this case. Combining them in the lower picture produces a much closer replica of the original surface, shown in the upper left-hand corner.

inconsequential. Computational efficiency is not an issue for a massively parallel brain with what may be, for all computational purposes, an unlimited number of logical units. We believe that the challenge of the computational load will ultimately be solved by the development of faster (and, most probably, parallel-processing) computers and, therefore, that it constitutes only a temporary technical problem.

The existence proof exhibited by the enormously powerful human visual system suggests that the combination approach that we and others have followed is likely to be the procedure of choice in the future. Although it is unlikely that the specific computational algorithms that have been used here are actually executed by the human nervous system, current experimental outcomes in the

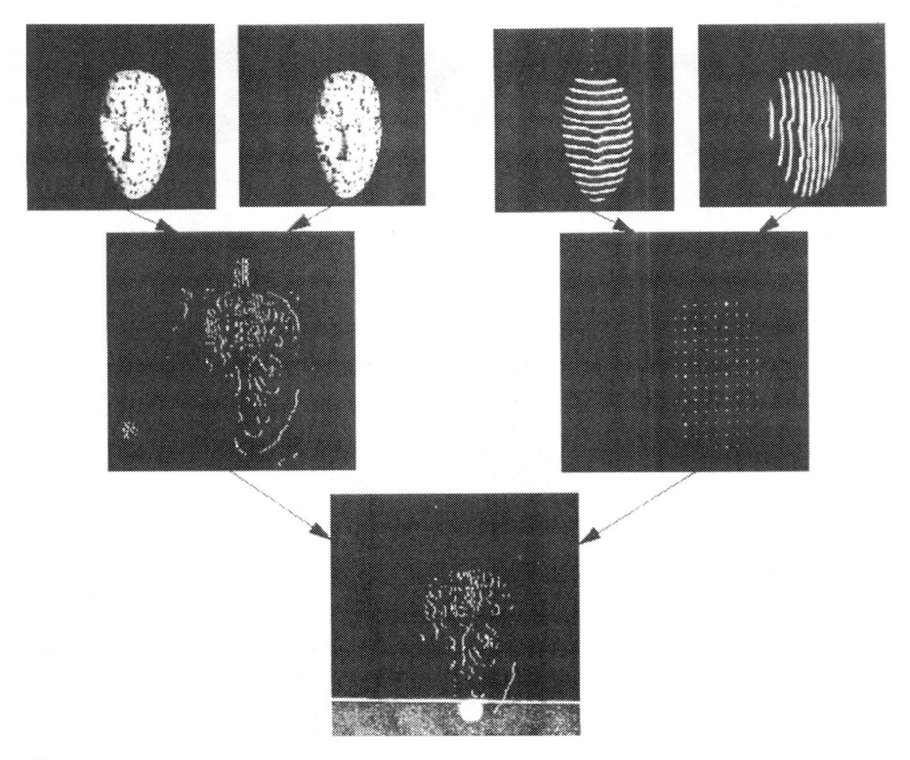

Figure 6.11 Diagram showing the functional interactions between the Stereo and SFSL algorithms. The SFSL constrains the algorithm to use only the needles in the region defined by the mask. The needles in this region are averaged to produce an estimate of the range from the cameras to the mask. Stereo is inadequate to reproduce the shape of the face, and range measures are contaminated by extraneous needles outside the mask region. SFSL is inadequate to measure the range to the mask. Together, however, they produce a good estimate of the range to the mask.

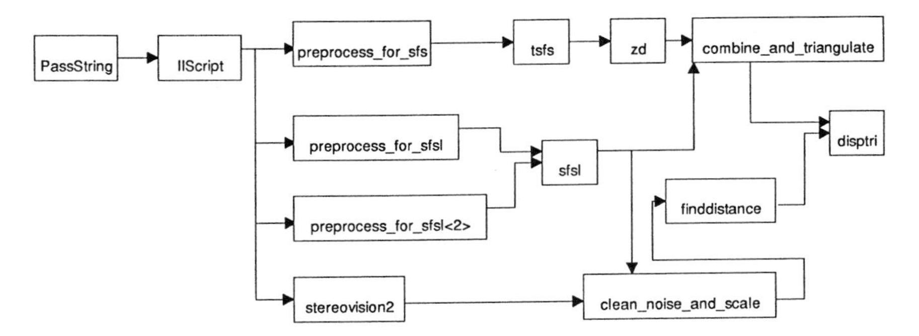

Figure 6.12 Map of the program that combines SFS and SFSL and SFSL and Stereo, respectively. The process operates as an integrated whole to produce better estimates of the range and the details of the surface shape than any one of the three could if used individually. This map is a model of organic spatial vision. Boxes marked *tsfs* and *zd* are the SFS operations. "PassString" and "IIScript" are housekeeping functions. Other components are described in the text.

field of visual perception and computer vision suggest that the general notion of interactive combination of comparable or analogous quasi-independent processes must underlay the way humans perceive three-dimensional scenes.

NOTES

1. This chapter is adapted from Uttal et al. (1996) and is used with the permission of the publisher.
2. We are grateful to MaryLou Cheal and Takeo Watanabe for helpful suggestions in the preparation of this chapter.

7

Object Recognition[1]

7.1 INTRODUCTION

Shape recognition is one of the most important steps in our model of a vision system. All of the segmentation and partitioning steps we have previously described are but preliminary to the central cognitive act of identifying the object—in other words, the process of recognition. Simply to be able to define contours or to identify the presence of an object would be of little adaptive value to the organism. Only when the organism can specifically identify an object can adaptive responses be selected and executed.

Object recognition, therefore, has always been a central and important part of many computational models of vision. Numerous mathematical, computer, and conceptual models have been developed to describe object recognition. Comprehensive reviews of object recognition (or, as it is often otherwise known, pattern or shape recognition) can be found in Grimson (1981) and Carpenter and Grossberg (1991) and in many other books on biological perception and computer image processing.

More specifically, Hoffman and Richards (1984) suggested that visual stimuli are broken down into simple two-dimensional parts for recognition. They proposed that it is the spatial arrangement of two-dimensional shape components that is the basis of visual recognition. Beiderman (1987) also suggested an account of human visual object recognition, in which visual objects were first decomposed into sets of idealized geometrical forms he called *Geons*. Geons, such as cylinders, spheres, and generalized conic sections, are invoked as the basic components of three-dimensional objects. In this account, the spatial arrangement of these geometrical primitives is also used to identify objects in three-dimensional scenes. Computational approaches to visual object recognition that operate on similar principles have been proposed by Henderson and Anderson (1984), Wojcik (1984), and Wu and Stark (1985).

This chapter describes a computational model of visual object recognition that is embedded in our SWIMMER vision system. Unlike many previous efforts, it is based on the invariant encoding and comparison of two-dimensional contours. By *invariant* we mean that the encoded representation of an image once it has been segmented from the background is the same regardless of its size, orientation, or position in the visual field. Furthermore we have designed our system so that it will be "invariant," to the maximum degree possible, with regard to objects that are partially occluded by others. The approach taken here also differs from previously described models in that it does not employ pre-existing shape primitives as the basis of visual recognition. Rather, the entire shape of the object is the primitive. We are convinced that this is a much more realistic model of human object recognition, since it mimics the human's sensitivity to the overall arrangement, or *Gestalt*, of the object. Models based on shape primitives or subcomponents are artificial, in the sense that there is little evidence that such subcomponents are actually processed during human recognition.

Instead of shape primitives, the present object recognition model is based on a representation of the entire outline of a object. It segregates an input scene into the constituent shape contours. A *shape contour*, in this context, refers to a two-dimensional curve produced by the isolation of the edge or boundary of an object. The shape contour may be complete, or it may represent only a fragment of the entire outline of the object. A *scene* in the following discussion is made up a number of shape contours, each of which represents one or another object in that scene. We also use the word *subcomponent* to denote either a partial or a complete shape contour of an object.

The recognition process in this part of our model is carried out by comparing each of the shape contours with others prestored in a library of shapes. A *recognition* is designated when the input shape contour matches one item in the library of shape contours. A *match* in this case occurs when the difference between the input and the library item is less than an arbitrary threshold.

Our model measures differences between the contours of arbitrary shapes by first encoding them into a position-, size-, and orientation-invariant representation. The key to the recognition model, therefore, is the measure of difference between the test object and the set of shape contours in memory. The difference between shapes is represented as a set of values describing local differences between two shapes. This set of difference values is complete, in the sense that it provides enough information to transform one shape into another or to identify one as a part of the other even if other parts are missing or if the two shapes are not exactly the same.

One of the most important attributes of our model, therefore, is that it is able to measure differences between groups of overlapping, incomplete, and occluded shape contours. This is accomplished because occluded parts of the object can be matched with complete shape contours in the library. The combined difference measure is real-valued and varies continuously with changes in shape. That is, unlike a measure that is Boolean or absolute (i.e., same versus different) or qualitative (e.g., square versus round), this measure describes how much one shape must be changed to produce the other (e.g., how much one must be bent to become identical to the other.)

Unlike discrete-set theoretic models that measure similarity in terms of shared subcomponents, (Geyer and DeWald, 1973; Tversky, 1977), this approach allows the comparison of figures that do not possess exactly the same subcomponents. Similarity, rather than identity, is all that is needed for a match, and the closeness of the similarity can be regulated by judicious choice of threshold parameters. The model is therefore able to compare geometrically distorted or noisy shapes that "look like each other" without being identical and to generalize over a range of visually similar shapes.

It is well known that human observers have the ability to recognize partially occluded and distorted objects at various positions, sizes, and orientations. This process is known as *stimulus generalization.* Since our model was designed to perform the same task, it was developed to agree as closely as possible with human psychophysical findings. Although it is not currently possible to account for the specific neural mechanisms of stimulus generalization in human vision, it is possible to describe some of the transformations performed by the visual system in carrying out recognition tasks.

For this reason, our goal in building this model was to identify ways in which the same information available to a human observer could be used to perform the same task on a computer. The task, in this case, is to compare, identify, and categorize shapes under conditions of shift, magnification (dilation), rotation, and occlusion. Our model, therefore, is explicitly a mathematical tool that we hope will help further the understanding of the shape recognition process in the human visual system.

7.2 THE ORGANIZATION OF THE MODEL

Any computational model of any process in visual perception must of necessity be constructed as a set of procedural modules. We simply do not know the organization of the whole system well enough to attempt a unitary analysis. This current version of our object recognition process model is divisible into four component procedures. The first of these procedures is a relatively standard application of edge enhancement techniques designed to segment and identify the contours of objects contained in a scene. The second procedure is critical to our model and novel. In this step the object contour is encoded in an invariant form.

The third procedure is the one in which the comparisons are carried out between a segmented object contour and the library of other forms coded in their own invariant representations. Finally, in the fourth procedure the specific process of categorizing and, thus, recognizing the form as an exemplar of one of the prototypes from the library is carried out. When combined, these procedures allow the model to analyze a visual image containing multiple occluded objects, identify familiar objects, and categorize unfamiliar objects. Each of these processes will now be described in detail.

7.2.1 Subcomponent Extraction

Subcomponent extraction in our model operates on the edge contours extracted from the objects in a scene. Edges in an image are locations where intensity changes sharply in any direction. The importance of edge information for human visual recognition has long been appreciated by artists and has been experimentally emphasized by recent psychophysical work, such as that reported by Beiderman and Ju (1988). Psychological models of edge enhancement and extraction have previously been described by Graham (1980), Morrone and Burr (1988), Tolhurst (1972), and Uttal (1981). Lateral inhibition in a horizontal array between neighboring receptors in either the invertebrate or the vertebrate eye (Hartline, 1949) also performs this same edge enhancement. Lateral inhibition increases sensitivity to localized intensity changes and decreases sensitivity to homogeneous regions in the biological system.

A wide variety of computational operators can be used to enhance or extract edge information from continuous, multilevel color or gray-scale images. Operators based on the first and second spatial derivatives of image intensity are commonly used to enhance edges. One example is the Laplacian of the Gaussian described by Marr and Hildreth (1980), a method that enhances the zero crossings of an intensity function, the same process used to establish correspondence between stereoscopic images as described in Chapter 6. The Laplacian of the Gaussian is also an edge enhancer, but it has the added advantage of providing additional information concerning the direction from which the zero-intensity contour was crossed as described there. This function may be expressed as:

$$V^2G = \frac{\delta^2 G}{\delta x^2} + \frac{\delta^2 G}{\delta y^2}$$

Many other specialized computational methods have been developed to enhance edges that are not formally derivative-evaluating algorithms. Nevertheless, these computational algorithms have the same sensitivity to regions of rapid change of intensity or color using pixel manipulations in a local region of the image. Algorithms suggested by Roberts (1965), Sobel (1970), Prewitt (1970), and Canny (1986) are among those most widely used currently. Other, more elaborate computational methods for edge enhancement and extraction have been proposed by Argyle (1971), Burns et al. (1986), Davis (1975), Haralick (1980), and Hashimoto and Sklansky (1987). Methods for thresholding and then thinning edges have been described by Deutsch (1972), Kong and Rosenfeld (1989), and Sahoo et al. (1988).

The effect of any of these procedures is to enhance edges or discontinuities and to diminish differences between homogeneous regions in an image. The product of any edge enhancement and extraction algorithm resembles a line drawing of the scene, an example of which is shown in Fig. 7.1.

In the present model, all shape contours consist of sets of Cartesian coordinates (x_i, y_i). These contours are obtained from the raw edge-enhanced information by sequential thresholding, dilation, and skeletonizing or erosion. This se-

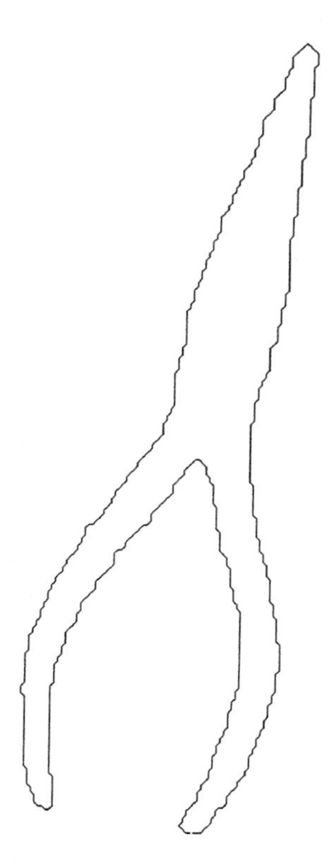

Figure 7.1 The edge pattern produced by processing a photograph of a pair of needle-nosed pliers.

ries of processes ultimately results in a shape contour of unit pixel width. Thresholding serves to suppress weak edges and to maximize the strong edges in an image. The superthreshold edges are then expanded by local dilation to eliminate small errors and gaps. These dilated edges are then thinned to the desired unit pixel width by a process of erosion or skeletonizing. The tables of numbers tracking the path of this shape contour make up the raw material for the next steps in our object recognition process. Contours extracted in this way define the shapes of any of the segmented objects in the original scene edges, but are more amenable to computational manipulation than would be the wider boundaries obtained from the raw-edge detection algorithm.

There is another problem that has to be faced at this point. So far we have not discussed in any detail the problem of objects that are overlapping—a situation for which the entire shape contour would be incomplete for some objects. In other words, some objects may be partially occluded. Thus, some of the shape contours may intersect with others, producing junctions where the occluding boundary of one object terminates the occluded boundary of another. In general, such junctions provide a powerful clue that at least one of them terminates abruptly. T-shaped boundaries represent a trivial and special case and are not required in our model; any three-way or greater junction is sufficient. The shape contours, whether they be terminated linear strings representing parts of an occluded object or a closed string representing the entirety of an unoccluded, and therefore closed, shape contour can then be dealt with as the subcomponents of the original scene.

The extraction of these shape contour subcomponents is accomplished by breaking the thinned edges at any points of intersection. Points of intersection are defined for any pixel whose connectivity is greater than 2. Pixels whose connectivity is less than 2 mark the terminating ends of a string—an isolated sequence of pixels. The points in a string are then ordered to form a connected sequence. The unoccluded portion of the occluded objects may be represented by these subcomponents, as shown in Fig. 7.2.

Connectivity is determined by the proximity of edge points in the image. A point (x,y) is connected to all other points (x_j,y_j) satisfying the condition represented by the following expression:

$$\sqrt{(x_j - x)^2 + (y_j - y)^2} \leq C \wedge$$

$$\neg\exists(x_k,y_k)\left[\sqrt{(x_k - x)^2 + (y_k - y)^2} < \sqrt{(x_j - x)^2 + (y_j - y)^2}\right.$$

$$\left.\wedge \tan^{-1}\left(\frac{x_k - x}{y_k - y}\right) - \tan^{-1}\left(\frac{x_j - x}{y_j - y}\right) \leq \theta\right]$$

where C and θ are the thresholds we set for separation and collinearity, respectively.

These equations indicate that two points are connected if and only if they are within distance C of each other and there is no other point that lies between them within an angle of size θ. Because this criterion for connectivity is real-valued, it is independent of the image coordinate system. These subcomponent contours of the original image, represented as a set of strings of pixel locations, are then selected so that they can be parameterized in terms of a single variable without violating continuity:

$$X(t) = x_i \; Y(t) = y_t$$

The result is a set of shape contours of unit width that have been isolated from any other contours with which they may have overlapped. This is the raw material for all subsequent image processing—the segmented shape contours.

Other methods for visual subcomponent extraction have been proposed by Baruch and Loew (1988), Fischler and Bolles (1986), and Rearick et al. (1988). Psychological descriptions of subcomponent grouping may be found in Hochberg and McAlister (1953), Kanizsa (1979), Koffka (1935), Kubovy and Pomerantz (1981), Ratoosh (1949), and Wertheimer (1938).

7.2.2 Subcomponent Encoding

The sets of pixels representing the extracted shape contours of a scene are then encoded to produce a position-, size-, and orientation-invariant representation. The initial representation is a normalized set of curve tangents describing the slope of the shape contour at each point as an angle relative to the vertical, as shown in Fig. 7.3. This angular normalized representation is then further normalized with respect to arc length, L_a, and average orientation, θ_a, in the following way:

$$\theta(t) = \theta(p) - \theta_a \qquad 0 \le t \le 1 \qquad tL_a = \int_0^{\rho} \sqrt{(X'(s))^2 + (Y'(s))^2}\, ds \qquad 0 \le \rho \le n$$

$$\theta(\rho) = \tan^{-1}\left(\frac{Y'(\rho)}{X'(\rho)}\right) \qquad 0 \le \rho \le n$$

$$L_a = \int_0^n \sqrt{(X'(s))^2 + (Y'(s))^2}\, ds \qquad 0 \le s \le n$$

$$\theta_a = \frac{1}{n}\int_0^n \theta(s)\, ds \qquad 0 \le s \le n$$

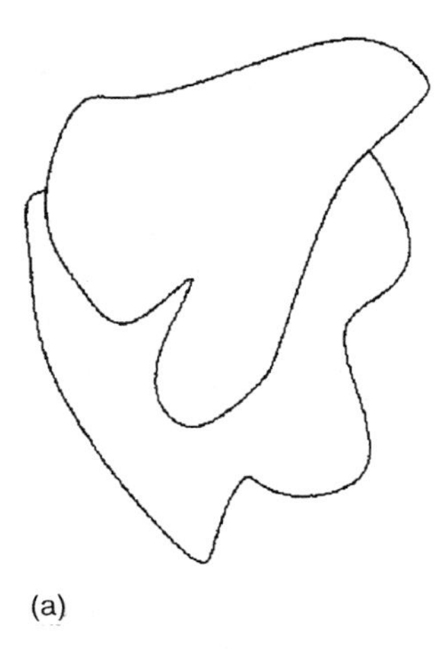

(a)

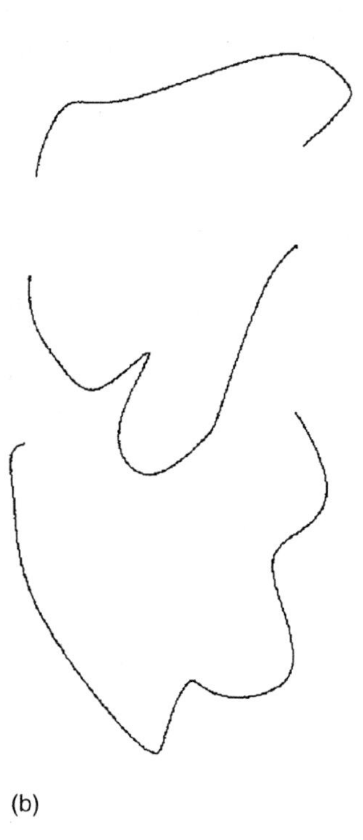

(b)

Figure 7.2 An example of feature extraction. (a) A scene showing two occluded objects. (b) Features isolated from the scene.

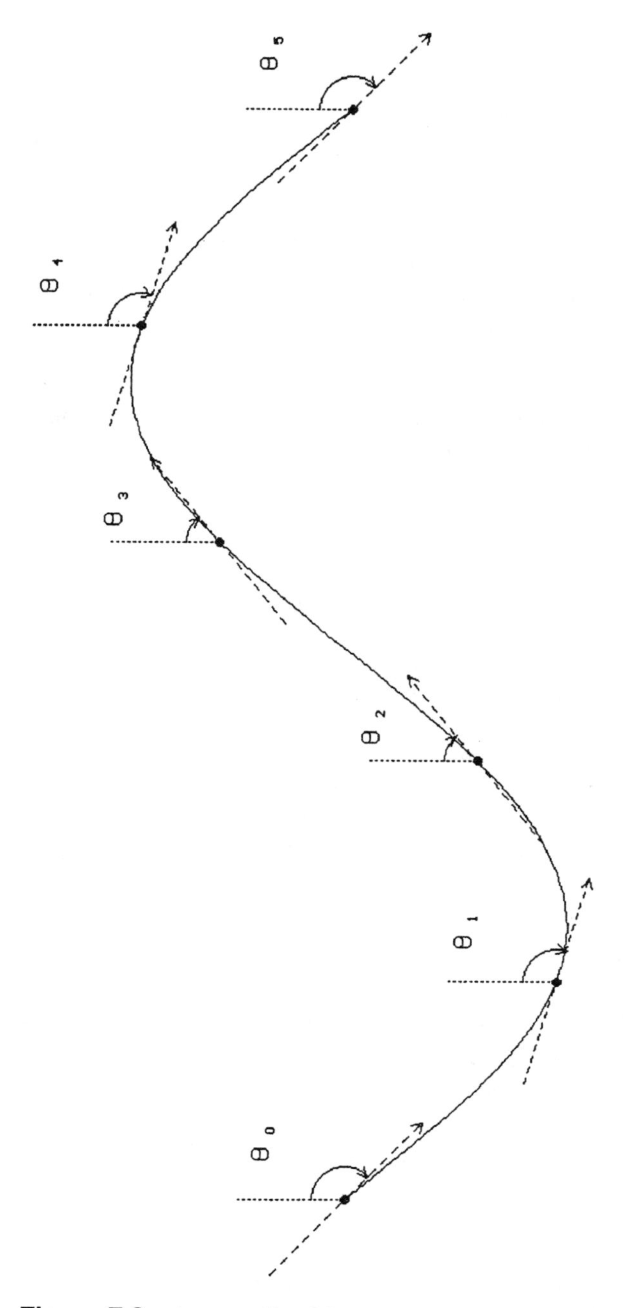

Figure 7.3 An example of feature encoding showing the tangent angles along a sine curve.

where n is the number of points in the contour string, t describes the position of a pixel along the contour as a function of its total arc length, $\theta(\rho)$ is a tangent to the contour at position ρ, θ_a is the average of the contour's tangents, L_a is the contour's total arc length, and $\theta(t)$ is the normalized tangent angle at position t. $\theta(\rho)$ is position invariant and is normalized to make it size and orientation invariant as well.

Size invariance is achieved by normalizing the contour in terms of its arc length. That is, each value of $\theta(t)$ is referenced by the proportion of the contour lying on either side of t. Thus, a point lying one-fourth of the way along the contour would be referenced by the parameter value 0.25, regardless of the contour's actual size.

To achieve orientation invariance, the mean of $\theta(\rho)$ is subtracted from each value of $\theta(\rho)$, normalizing it to have a mean of 0. Linear interpolation is applied to this representation to compensate for uneven spacing. The interpolation results in an evenly spaced distribution of tangent values. The representation $\theta(t)$ is position, size, and orientation invariant and can be compared directly with the representation of any other contour.

W. C. Hoffman (1966, 1980) describes such invariants in terms of the Lie group transformations of shift, rotation, and dilation. Attneave and Arnoult (1956) proposed a method for the invariant encoding and measurement of visual forms based on edge contours. Studies by such researchers as Bundesen and Larsen (1975), Cooper (1975), Cooper and Shepard (1973), Metzler and Shepard (1974), Sekuler and Nash (1972), and Shepard and Metzler (1971) suggest that perceptual invariance results from active stimulus transformations during visual processing. Computational and mathematical approaches to visual encoding have been described by Fitzpatrick and Louze (1987), H. Freeman (1960), Gray (1971), Li and Zhu (1988), Nalwa and Pauchon (1987), Wojcik (1984), and Zucker et al. (1988).

7.2.3 Subcomponent and Object Comparison

The invariant representations of the subcomponent shape contours extracted from an image can be compared by subtracting one from the other as follows:

$$D_{ij} = \int_{0}^{1} |\theta_i(t) - \theta_j(t)| \; dt \qquad 0 \le t \le 1$$

The total difference between two subcomponents, as described by this equation is the sum of the absolute differences between their corresponding tangent values. Each invariant representation is parameterized in terms of values between 0 and 1. Values of this parameter therefore refer to two corresponding tangent values, one from each subcomponent.

For practical reasons, the number of parameter values used in the computer implementation is limited to 100. This number of parameter values is almost always enough to compare accurately any two subcomponents. The resulting disparity measure describes how much the invariant representation of one subcomponent would have to be changed to produce the other. This measure describes the total difference between the shapes of the subcomponents.

To handle patterns that are only partially visible or occluded, the best match must be found between a partial subcomponent and a complete one by evaluating the following expression:

$$D_{ir} = \min_{0<k\leq 1, 0\leq l\leq 1-k}\left(\int_0^1 |\theta_i(t) - \theta_j(kt + l)|\ dt\right)$$

This is achieved by repeatedly sampling and comparing the subcomponents. The samples must vary in both size and location within the subcomponent. This is necessary because it is impossible to predict what portion of a subcomponent is visible in an occluded scene. Although the number of possible samples is very large, if they are evenly distributed over the whole contour, a smaller subset of the possible samples is generally enough to identify the subcomponent.

The difference between two whole patterns is calculated by averaging the differences between their respective subcomponents according to the following equations:

$$D_{p1,p2} = \frac{\Sigma \min_{ij}(D_{ij})}{\min(m,n)}\ 1 \leq i \leq m 1 \leq j \leq n$$

$$P_{p1,p2} = \frac{\Sigma P_{ij}}{\min(m,n)}\ 1 \leq i \leq m 1 \leq j \leq n$$

$$P_{ij} = (P_{xi} - P_{xj},\ P_{yi} - P_{yi})$$

$$(P_{xi}, P_{yi}) = \left(\int_0^1 X_i(t)dt,\ \int_0^1 Y_i(t)dt\right) 0 \leq t \leq 1$$

$$S_{p1,p2} = \frac{\Sigma S_{ij}}{\min(m,n)}\ 1 \leq i \leq m 1 \leq j \leq n$$

$$S_{ij} = \frac{L_{a_i}}{L_{a_j}}$$

$$\theta_{p1,p2} = \frac{\Sigma\theta_{ij}}{\min(m,n)} \quad 1 \leq i \leq m \; 1 \leq j \leq n$$

$$\theta_{ij} = \theta_{a_i} - \theta_{a_j}$$

$$\neg\exists(k,l)[(k = i \wedge l \neq j) \vee (l = j \wedge k \neq i)]$$

Here, $D_{p1,p2}$ is the total difference between the shapes of two patterns, and it is equal to the sum of the minimum differences between their respective subcomponents. $\mathrm{Min}_{ij}(D_{ij})$ is the smallest shape disparity between subcomponent i of one object and any subcomponent j of the other. $P_{p1,p2}$ is the total difference in the positions of two patterns and is equal to the mean of the differences in the positions of their respective subcomponents. P_{ij} is the difference in position between subcomponent i in one object and its best matching counterpart, subcomponent j, in the other. The position of a subcomponent is equal to the mean of its x- and y-coordinates. $S_{p1,p2}$ is the ratio of the sizes of two patterns. It is equal to the mean of all S_{ij}, the ratios of the sizes of their respective subcomponents. S_{ij} is equal to the ratio of the arc lengths of subcomponent i and its best matching counterpart, subcomponent j. $\theta_{p1,p2}$ is the total orientation difference between two patterns and is equal to the mean of all θ_{ij}, the orientation differences between their best matching subcomponents. θ_{ij} is equal to the orientation of subcomponent θ_i minus the orientation of the best matching subcomponent, θ_j. The final equation in this set indicates that no subcomponent in one object may be matched to more than one subcomponent in the other in any of the previous equations. Taken together, these equations describe the overall differences between the shapes, positions, sizes, and orientations of two patterns composed of multiple subcomponents.

Other approaches to measuring visual similarity have been described by Goldmeier (1972). Set theoretic accounts of similarity, based on shared subcomponent sets, have been proposed by Geyer and DeWald (1973) and Tversky (1977). Multidimensional scaling approaches to similarity measurement are described in Attneave (1950), Krumhansl (1978), Nosofsky (1985), Schonemann et al. (1985), Shepard (1957), and Torgenson (1965). Computational approaches to visual comparison are described in Maragos (1988) and Price (1985).

7.2.4 Hypothesis Testing and Verification

Once the comparison of the segmented and normalized shape contours has been carried out, it is necessary to categorize computationally and thus to recognize the objects that were part of the original scene. The matching process is complete at this point. This is accomplished by comparing each subcomponent in an incoming object with all subcomponents of all patterns in memory. Each comparison produces a shape disparity, D, a position difference, P, a size ratio, S, and an orientation difference, θ, for the specific patterns and subcomponents under comparison. The best matching set of subcomponents from memory is used to identify the incoming pattern.

Each of these subcomponent matches is treated as a hypothesis that a specific object is present in the image at a particular position, size, and orientation. To determine which patterns in memory best account for the incoming scene, each potential match is tested against the image. Patterns from memory are transformed to the position, size, and orientation indicated by each potential match and compared with the image.

In this secondary comparison, the image is checked for points that most nearly match those in the memory object:

$$\forall i \in P_1, \exists j \in P_2 \left[\sqrt{(x_i - x_j)^2 + (y_i - y_j)^2} < 20.0\right] \Rightarrow j \in P_1$$

This equation states that for all points in the memory object, if a point exists in the image that falls within 20 units of it, that point is recorded along with the subcomponent it belongs to. If the memory object accounts for at least 90% of a subcomponent in the image, that subcomponent is recorded as belonging to the pattern.

The memory pattern that accounts for the largest number of subcomponents from the image is recorded and its corresponding subcomponents removed from the image. In this way, a scene with multiple occluded objects is parsed into the set of memory objects that account best for its subcomponents. Any subcomponent that is unaccounted for at the end of this process is considered ambiguous and is later added to memory as part of a new object.

Thus, identifying objects in the image is a two step process. In the first step, hypotheses are formulated about what could be in the image based on an invariant comparison of partial subcomponent information. In the second step, these hypotheses are tested using a noninvariant comparison with a complete pattern from memory. This two-step process takes both local invariant and global noninvariant information into account, making it more efficient and more accurate than either process alone.

Because the invariant representation is derived from the first derivative of the contour, it is locally position invariant but highly sensitive to curvature. As a result, it may tend to favor shapes that are similar in local curvature, even if they are not spatially congruous. This makes it difficult for the invariant comparison to determine the global congruity of an object that is divided into several smaller parts. The resulting trade-off is one of more efficiency for less accuracy. By testing the results of the invariant comparison with a complete two-dimensional representation, accuracy is greatly improved, with only a minor decrease in efficiency.

Treisman (1986a) suggests that human vision may be divisible into similar processing stages. The early stages are thought to be highly efficient at detecting patterns but poor at spatial localization. This is evidenced by illusory stimulus conjunctions that violate spatial contiguity. Later stages of processing that are more sensitive to spatial localization, if uninterrupted, prevent the emergence of these illusory conjunctions.

Once the objects in a scene have been identified, the scene is reconstructed using information in memory. Each of the identified patterns is drawn from memory and placed in the scene at the position, size, and orientation indicated by the comparison process. Parts of shapes that are occluded in the original image may be color coded to illustrate the spatial relationships of the objects. The reconstructed scene is an interpretation of the original, which contains additional information about the true shapes of objects and their spatial arrangement. Objects in the scene are also presented in isolation, to illustrate further the model's interpretation.

7.2.5 Categorization of Forms

In addition to matching objects in a scene with objects in memory and reconstructing the scene, the model includes an associative memory for object attributes. This associative memory uses correlational links between objects in memory and categorical information describing them. The descriptive information may include names, adjectives, functions, related objects, and contexts. In addition to providing information about familiar objects in a scene, the associative memory is used to categorize unfamiliar objects by forming new associative links with existing descriptors in memory.

The strengths of associative links between objects in memory and their descriptors are determined by correlation. Occurrences of a shape are correlated with a set of descriptors provided as teaching inputs. The strength of an association between a pattern in memory and a descriptor is calculated as:

$$A_{ij} = \frac{1}{n}\sum_{k=1}^{n} d_{jik} \qquad d_{jik} = [0,1]$$

where A_{ij} is the strength of the association between a object denoted by i and a descriptor denoted by j. The variable d_{jik} indicates the presence (1), absence (0), or negation (−1) of descriptor j with the kth presentation of object i out of the n most recent presentations. The more often a descriptor occurs as a teaching input with the presentation of a specific pattern, the stronger its associative link with that pattern.

Descriptors associated with an object in memory are recalled according to a weighting function that allows the model to generalize and degrade gracefully. This weighting function calculates association strengths between a new object and the descriptors in memory, as follows:

$$W_{ij} = \sum_{k=1}^{n} \frac{\left(\prod_{l=1, l \neq k}^{n} D_{pipl} \right) A_{kj}}{\sum_{m=1}^{n} \left(\prod_{l=1, l \neq m}^{n} D_{pipl} \right)}$$

The resulting weight value represents the likelihood that the incoming object, denoted by i, has the attribute denoted by j, based on its similarity to objects, k, in memory.

If a perfect match is found between the incoming object, denoted by i, and one in memory, k, then the disparity, D_{pipk}, is equal to zero. The zero value of D_{pipk} for the matching object makes the value of

$$\left(\prod_{l=1, l \neq m}^{n} D_{pipl} \right)$$

equal to zero for all cases except $m = k$. In this case, the weighting function reduces to $W_{ij} = A_{kj}$ and the associative links of the matching object in memory are directly inherited by the incoming object. This is an example of *nearest-neighbor categorization*, in which the incoming object inherits the categorical attributes of the closest matching exemplar in memory.

If the incoming object is equally different from all objects in memory, the values of

$$\left(\prod_{l=1, l \neq k}^{n} D_{pipl} \right) A_{kj}$$

are equal for all k. As a result, the values of W_{ij} are equal to the arithmetic mean of the association strengths A_{kj}. That is:

$$W_{ij} = \sum_{k=1}^{n} \frac{A_{kj}}{n} \qquad \text{for all } k$$

This is an example of *prototype categorization*, in which the incoming object inherits the attributes of the category prototype. A category prototype is generally defined as the most central member of the category. Its attributes are the average of all category member's attributes.

If the differences D_{ij} between the incoming object and objects in memory are not equal and are nonzero, the values of W_{ij} are equal to the weighted means of the association strengths A_{kj} of all objects in memory. This means that the incoming object inherits categorical information from all objects in memory in an amount proportional to their similarity. Thus, the more similar the incoming object is to one in memory, the more categorical information it inherits from that object. This method of categorization has the advantage that it is able to handle familiar and unfamiliar objects equally well. Malt (1989) has demonstrated experimentally that human subjects are able to use both exemplar and prototype strategies within a single categorization task.

Our model is also similar to the following one proposed by Nosofsky (1988):

$$P(R_j | S_i) = \frac{\sum_{j \in C_j} N_j S_{ij}}{\sum_k \left(\sum_{k \in C_k} N_k S_{ik} \right)}$$

where $P(R_j|S_i)$ is the probability that stimulus i will be placed in category C_j, N_j is the frequency of presentation for exemplar j in category C_j, and s_{ij} is the similarity between stimulus i and exemplar j. This equation states that the probability of an incoming stimulus being placed in a given category is equal to the weighted sum of its similarity to all exemplars within that category divided by the weighted sum of its similarity to all exemplars in all categories. The Nosofsky model has difficulty correctly handling specific category exemplars and exceptions. Unlike the present model, the Nosofsky model is influenced by all exemplars, even in the presence of a perfect match.

Because this approach to categorization uses an associative memory based on multiple descriptors, it can determine category membership in a very flexible manner. For example, if a category is defined by a set of descriptors, the association strengths can be collapsed over that set to determine category membership. Individual association strengths can be weighted in proportion to their correlation with the given category to produce an overall association strength for the category. Thus, a new category can be defined by its correlation with each of the descriptors in memory without requiring an explicit definition. As new contexts and new categories arise, the same set of association strengths can be used to make categorical decisions without retraining or explicit rules.

Various theories and studies of categorization have been described by Estes et al. (1989), Hock et al. (1987), Homa et al. (1991), Murphy and Wisniewski (1989), Nosofsky, Clark, and Shin (1989), Nosofsky (1988), Rosch and Mervis (1975), and Stromer and Stromer (1990a,b).

7.3 VALIDATING EXPERIMENTS

To test the model developed in the preceding sections we carried out two validating experiments. The first experiment tested the model's ability to segment and identify objects of varying size, position, and orientation. In other words, this was a specific test of the model's ability to ignore these parameters of an image. In this manner, we were able to compare its recognition skills with that of a generalized human observer.

The second experiment was designed specifically to compare our model with another, similar one proposed by Nosovsky (1988) to describe categorizing behavior.

7.3.1 Experiment 1

This experiment was designed to test the model's ability correctly to identify occluded and unoccluded objects with position, size and orientation invariance. Scenes containing 1–3 objects at randomly selected positions, sizes, and orientations were given as data to a computer program. This program calculated the invariant representations of the incoming forms and compared them with objects in memory according to the model.

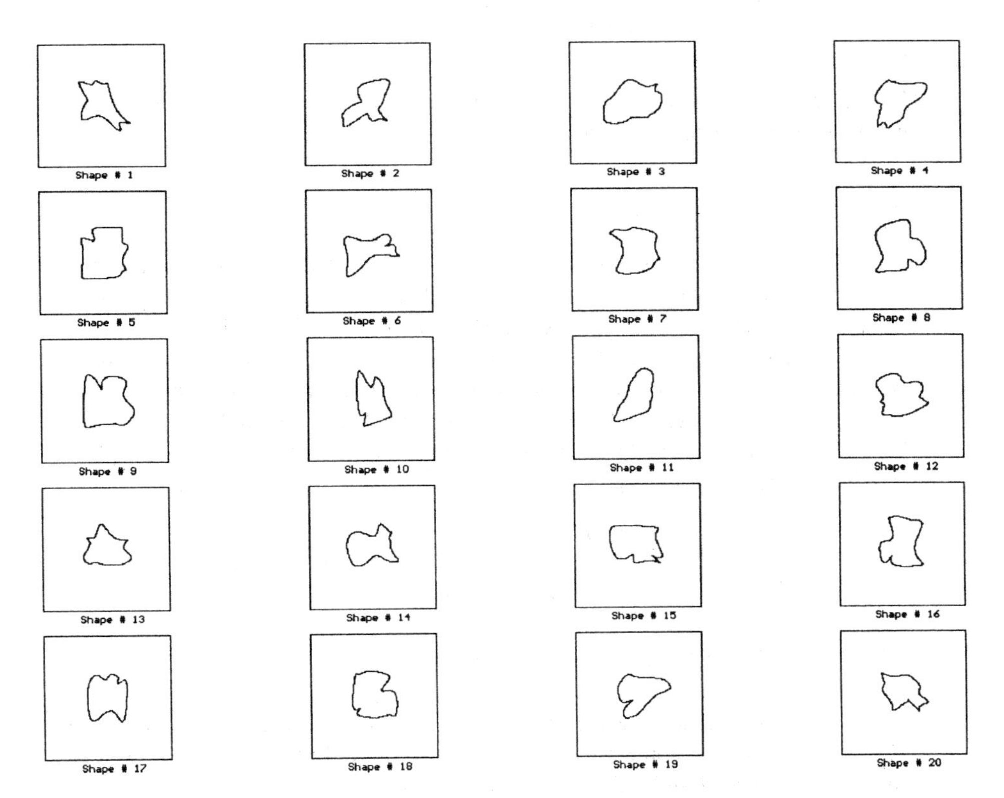

Figure 7.4 Twenty shapes in the library used for recognition in Experiment 1.

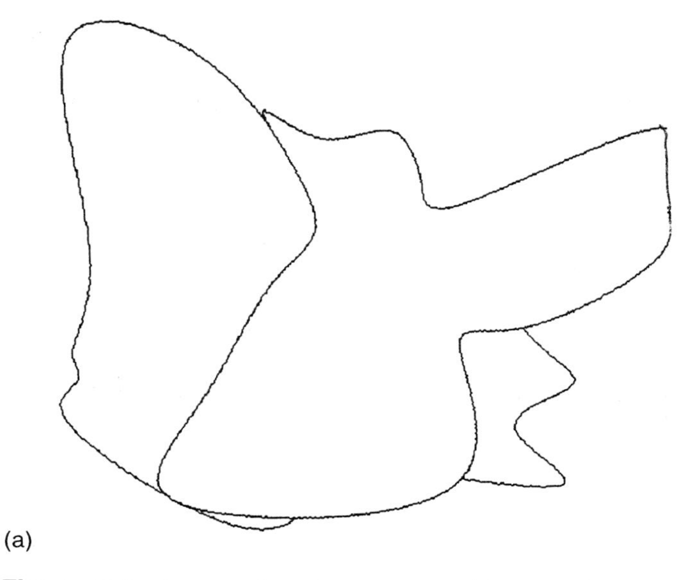

Figure 7.5 Interpretation of a scene containing two shapes. (a) The overlapping shapes. (b) The computer interpretation.

Method

Stimuli. Stimuli used in this experiment consisted of the 20 randomly generated, closed shapes shown in Fig. 7.4. These shapes were generated by randomly placing points within a 512×512 grid. The center of gravity of this set of points was calculated, and the points were converted to polar coordinates. The polar coordinates were then sorted in clockwise order to ensure that the shape would be a closed contour. All of the sorted points were then converted back to Cartesian coordinates and used as control points for a two-dimensional cubic *B*-spline. The *B*-spline was defined by the following parametric equation on *X* and *Y*:

$$Q_i(t - t_1) = (X(t - t_i), Y(t - t_i))$$

$$= \frac{(1-t)^3}{6} P_{i-3} + \frac{3t^3 - 6t^2 + 4}{6} P_{i-2} + \frac{-3t^3 + 3t^2 + 3t + 1}{6} P_{i-1}$$

$$+ \frac{t^3}{6} P_i \; P_i(x_i, y_i) \qquad 0 \leq t \leq 1.0$$

The resulting shapes are *C*2 continuous, meaning that all spline segments have the same first- and second-order terms and are smoothly joined. These shapes are smoothly curved and lack sharp discontinuities.

After the shapes were generated, they were added into a memory file used

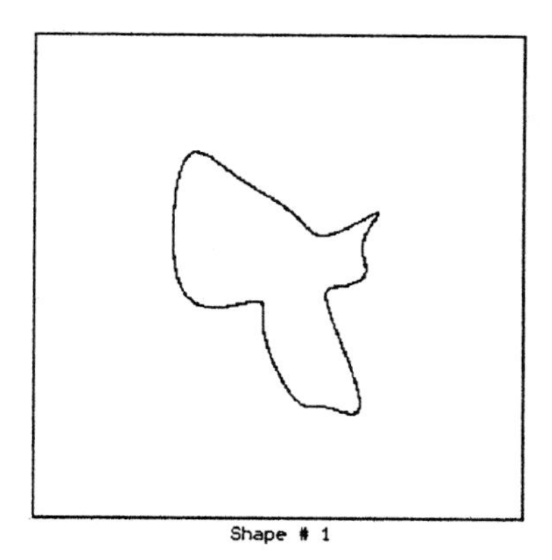

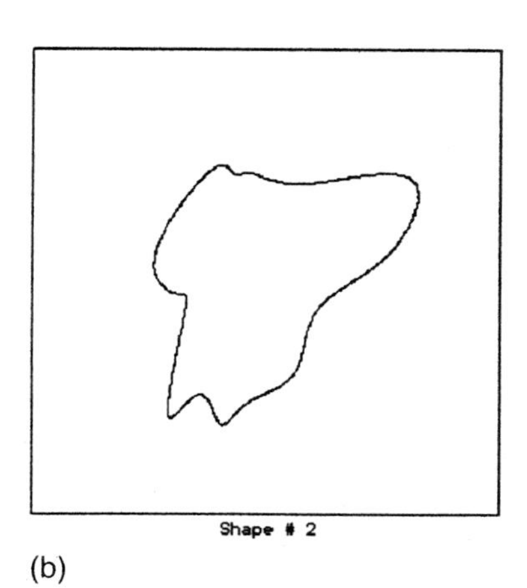

(b)

by the computer program. Next, individual shapes were combined into scenes of one, two, or three objects. Random shifts, rotations, and scalings were applied independently to each of the shapes in a scene as follows:

$$X_2(t) = ((X(t) - xc)\cos\theta + (Y(t) - yc)\sin\theta)s + xc + \delta x$$
$$Y_2(t) = ((Y(t) - yc)\cos\theta - (X(t) - xc)\sin\theta)s + yc + \delta y$$
$$xc = \int_0^1 X(t)\ dt$$

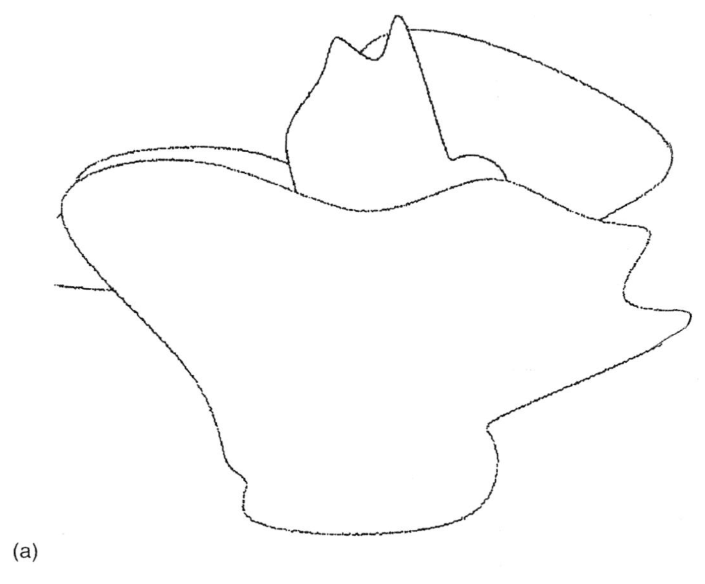

(a)

Figure 7.6 Interpretation of a scene containing three shapes. (a) The overlapping shapes. (b) The computer interpretation.

$$yc = \int_0^1 Y(t)\, dt$$

$$0 \leq t \leq 1$$

where $X_2(t)$ and $Y_2(t)$ are the parametric functions describing the transformed shape, (xc,yc) is the shape's center of gravity, θ is the angle of rotation, s is the scaling factor, and $(\delta x,\delta y)$ is the shift. Shifts on x and y were limited to ± 50, rotations were between 0 and 2π radians, and scale was between 1.0 and 0.5.

Once transformed in this way, the shapes were overlaid to produce 300 scenes of occluded objects. The shapes were always assumed to be opaque, such that each shape hid parts of the shapes behind it. Edges of the shapes that extended beyond the borders of a 646×486 rectangle were cut off as though the shapes were occluded by the viewing window of a computer CRT screen. Doing this further challenges the computer by limiting its field of view, producing a more robust test of recognition accuracy. These scenes were then given to the computer program for recognition.

Procedure. Scenes of one, two, or three objects were given as data to the computer program. The program then attempted to identify occluded and unoccluded objects in the scenes. Information about the identity of each object in a scene was returned along with its position, size, and orientation relative to the

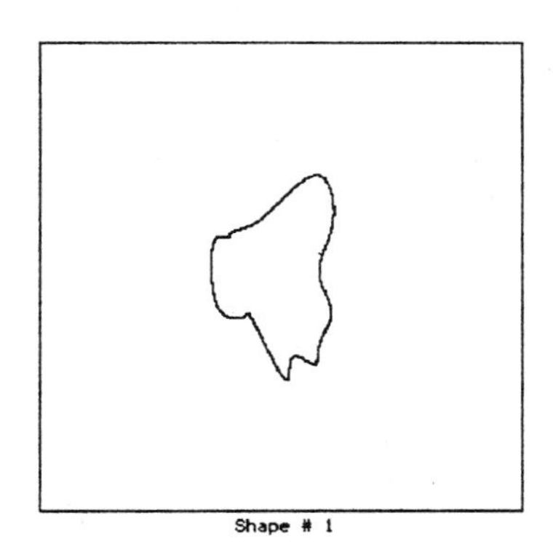

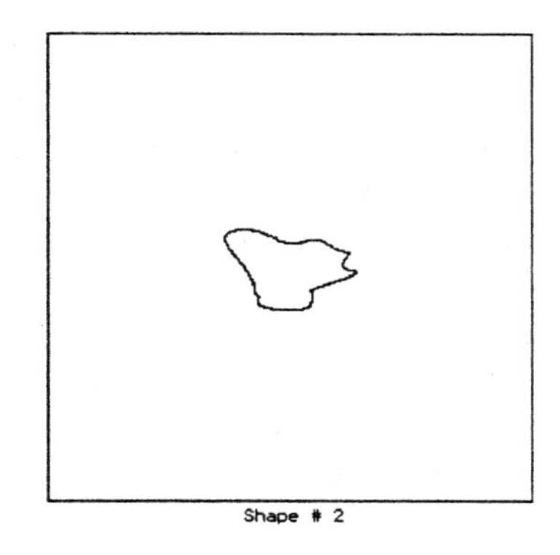

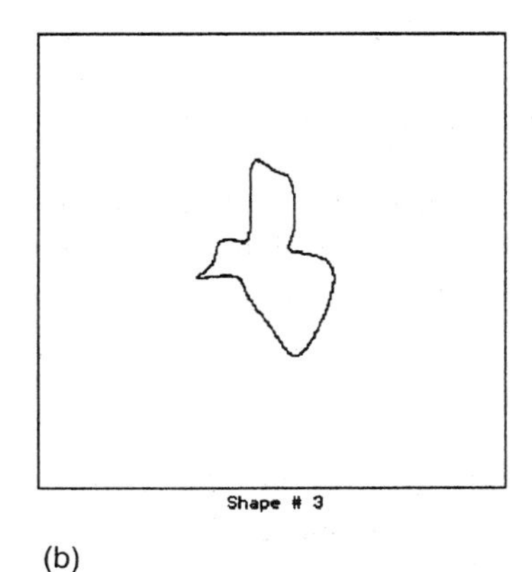

(b)

target object in memory. Hits, misses, and false alarms were recorded for each scene. An example of a scene containing two incomplete objects is given in Fig. 7.5. A corresponding example for an image consisting of three incomplete shapes is given in Fig. 7.6.

A hit was defined as the identification of an object in the scene such that the identified shift, rotation, and scale values for the object were all within a specified limit of their true values. The specified limits were 20 pixels for shift, 0.6 radians (18°) for orientation, and 0.05 for scale.

In the computer implementation, the invariant representation of each subcomponent was limited to 100 evenly distributed values out of a possible 1400. Oversampling during the first phase of the comparison process was limited to 40 different sizes and 40 positions, for a total of 1600 samples per object in memory out of a possible 1.96×10^6.

During the second phase of the comparison, the best 10 matches for each subcomponent in the scene were used for hypothesis testing out of a possible 3.92×10^7. The reason for these optimizations was twofold: First, they substantially decrease the running time of the computer program; second, they provide a way to determine the robustness of the recognition method.

Results. Results for scenes containing one, two, and three shapes are shown in Table 7.1. Chance recognition performance for a 1-of-20 forced-choice task, based on a library of 20 shapes is $P(\text{Hit}) = 0.05$, $P(\text{Miss}) = 0.95$. Recognition accuracies in the one-, two-, and three-shape conditions were significantly greater than predicted by chance, $\chi^2(1, N = 200) = 180.95$, $p < .00001$, $\chi^2(1, N = 400) = 211.74$, $p < .00001$, $\chi^2(1, N = 600) = 259.09$, $p < .00001$. Pairwise comparisons between these conditions indicated that recognition accuracy declined significantly between the one- and two-shape conditions, $\chi^2(1, N = 300) = 27.87$, $p < .00001$), and between the two- and three-shape conditions, $\chi^2(1, N = 500) = 3.94$, $p < .05$, although the difference between the two- and three-shape conditions was marginal. False alarm rates also increased between the one- and two-shape conditions, $\chi^2(1, N = 300) = 11.57$, $p < .0007$, and between the two- and three-shape conditions, $\chi^2(1, N = 500) = 15.17$, $p < .00011$.

These results demonstrate the accuracy and robustness of the model's shape-encoding and comparison processes. Even in the presence of high sampling error, mean recognition accuracies for scenes containing one, two, or three shapes were 100%, 76.5%, and 68.33%, respectively.

7.3.2 Experiment 2

This experiment was designed to compare categorizations produced by the Nosofsky model with categorizations produced by the present model. Both models were tested under two conditions. In the first condition, the models were tested for categorization of random deviates from a set of prototypes. In the second, the models were tested for categorization of random deviates from a set of exemplars. The same overlapping categories were used for both tests to ensure

Table 7.1 Recognition Performance for 1, 2, and 3 Shape Conditions

	Number of Shapes		
	1	2	3
Hits	100	153	205
Misses	0	47	95
False alarms	0	24	92

that category membership was not determined completely by distance from the prototype.

Method

Stimuli. The stimuli used in this experiment were 10 categories defined in two dimensions. These categories were generated by randomly placing 10 points in a 1024×1024 grid. Each of these points served as a category prototype. One hundred random deviates were then generated in polar coordinates around each of the 10 prototypes. These random deviates spanned a distance of 0–256 units from the prototype. The resulting groups of points formed partially overlapping, circularly symmetric categories.

Procedure. Both models were tested in a prototype condition and an exemplar condition. In the prototype condition, both models were used to categorize random deviates around each of the 10 category prototypes. The random deviates were generated in polar coordinates at distances of 0–256 units from the prototypes. The distributions of deviates were circularly symmetric around the prototypes. One hundred random deviates around each prototype were categorized using each of the two models for a total of 1000 test trials. Hits and misses were recorded for each of the models. A hit was defined as the model's favoring the category around whose prototype the random deviate was generated. A miss was defined as the model's favoring any other category.

In the exemplar condition, a single random deviate was generated in polar coordinates around each exemplar in each category. The random deviates ranged in distance from 0 to 10 units from the exemplars. The distribution of random deviates was circularly symmetric. A total of 1000 random deviates were categorized using each of the models. Hits and misses were again recorded for each model. A hit was defined, in this case, as the model's favoring the category around whose exemplar the random deviate was generated. A miss was defined as before.

Results. The critical factor in this experiment was how well the models handled both prototype and exemplar conditions. Hits and misses for each model in each condition are reported in Table 7.2. Chance performance in these experimental conditions is defined by a 1-in-10 forced choice as $P(\text{Hit}) = 0.1$. Each of the two models performed better than chance in the prototype condition, $\chi^2(1, N = 2000) = 864.45$, $p < .00001$, $\chi^2(1, N = 2000) = 859.67$, $p < .00001$. Performance of the two models did not differ significantly in this condition, $\chi^2(1, N = 2000) = 0.01$, $p > .9$.

Each model also performed significantly better than chance in the exemplar condition, $\chi^2(1, N = 2000) = 285.00$, $p < .00001$, $\chi^2(1, N = 2000) = 217.37$, $p < .00001$. The present model performed significantly better than the Nosofsky model in this condition, however, $\chi^2(1, N = 2000) = 5.60$, $p < .02$.

These results demonstrate that the present model is more effective at exem-

Table 7.2 Categorization Performance for Prototype-Based and Exemplar-Based Conditions

	Model	
Stimulus class	Present model	Nosofsky model
Prototype		
Hits	750	748
Misses	250	252
Exemplar		
Hits	434	382
Misses	566	618

plar-based categorization than the Nosofsky model, although it is equally effective at prototype-based categorization. This is important because it allows the model to recognize specific exemplars and also to categorize new stimuli. As a result, the model works better in situations where category boundaries are ill defined or outliers are present.

7.4 GENERAL DISCUSSION

The model described in this chapter is intended as a mathematical tool to account as accurately as possible for the invariant recognition and categorization of occluded and unoccluded objects in visual scenes. It demonstrates how edge contours in an image may be encoded and compared with representations in memory to identify objects and interpret visual scenes. An experimental test demonstrated that this encoding and comparison process was sufficient to identify objects in a position-, size-, and orientation-invariant manner. The model's ability to identify occluded objects and interpret scenes containing multiple objects was also demonstrated. Recognition performance of the model was also found to be robust with respect to large sampling error.

A mathematical account of categorization was also presented as part of the overall model. The model was shown to account for both prototype-based and exemplar-based categorization strategies. This allows it to handle both the recognition of familiar objects and the categorization of unfamiliar objects. In an experimental test, the performance of this model was compared with that of another current model of categorization, proposed by Nosofsky (1988). The present model matched the performance of the Nosofsky model for prototype-based categorization, but was significantly better at exemplar-based categorization.

Weaknesses of the present version of the model are also important to consider for its future development. The model's primary weakness stems from the nature of visual contours. For the model to operate correctly, edge contours must be reliably extracted from the image. While on the surface this appears to be fairly simple, it is extremely difficult to identify object boundaries and contours in an image reliably without prior knowledge about the visual scene.

Edge contours that are discontinuous or noisy greatly diminish the accuracy of contour grouping, encoding, and comparison. Although the model is relatively insensitive to low-frequency shape distortions, it is highly sensitive to erroneous contour groupings and high-frequency distortions. Much of the effect of high-frequency noise can be eliminated by a process of smoothing or filtering, but the result is an accompanying loss of fine detail.

One possible solution to this problem is to compare shapes in a way that does not depend on continuous contours. A point-to-point comparison that determines the minimum difference between shapes would serve this purpose. This solution would be independent of contour groupings and would not require the extensive preprocessing currently needed to extract edge contours. Using such an approach, the model would still be able to utilize edge contours, but it would not be limited to definable contours or contour groupings.

By gradually modifying and improving the present model and other models of visual object recognition, it will be possible to define better the relationship between the physical attributes of a shape and its perceptual and categorical identity. In this way, the fields of mathematical and computational modeling lend to the greater understanding of shape perception and psychophysics.

NOTE

1. This chapter is adapted from a dissertation submitted by T. S. Shepherd to the Department of Psychology at Arizona State University in partial fulfilment of the requirements for the degree of Doctor of Philosophy.

8

Surface Reconstruction[1]

8.1 INTRODUCTION

As we saw in Chapter 3, one of the most important parts of any visual system is the reconstruction function. Reconstruction may involve the filling in of missing parts ranging from major components of an otherwise complete image to the filling in of a sparse set of samples to make a complete surface. In Chapter 5, when we discussed our recognition programs, we saw how occluded and thus incomplete images were completed by using additional information obtained from a library of complete forms.

In this chapter we deal with another aspect of the reconstruction process: interpolation. Interpolation theory is a highly developed mathematical technique that has emerged into special importance in recent years as digital computers were developed to aid in the design of complex surfaces such as automobile bodies and machined parts. Interpolation problems also arise in fields as diverse as structural mechanics, electrostatics, computational fluid dynamics, and computer vision. The process of rendering sparsely sampled surfaces is also obviously an important part of visualizing any kind of sampled data and, as we see later, plays an important role in our interpretation of the results of our vision system. Our readers are specifically directed to an exactly analogous psychophysical experiment carried out earlier in our laboratory (Uttal et al., 1988).

In this chapter we present a method that would be especially useful to model and explain the perception of apparently complete surfaces from a relatively small sample of randomly scattered points. The need for interpolation of curves and surfaces to scattered data has been satisfied with a wide variety of mathematical techniques and tools, each of which has its advantages depending on the setting of the problem or the application for which it is intended. A subset of this wide-ranging problem involves fitting a surface defined by some function to the z-axis data f_i sampled at arbitrary locations in a planar domain—the x,y projection of the three-dimensional surface to be reconstructed.

Stated formally, the problem of bivariate scattered-data interpolation can be described as follows. Given n distinct arbitrary points (x_i,y_i) in the plane and n real values f_i associated with each of these locations, the objective is to construct a smooth function $F(x,y)$ that satisfies $F(x_i,y_i) = f_i$ for all $i = 1, \ldots, n$. This problem has been studied extensively, and a number of survey papers (Barnhill, 1977, 1985; Franke, 1982a, 1987; Franke and Nielson, 1991; Nielson, 1993; and Schumaker, 1976) provide a good review of some of the available techniques. The process in general is referred to as *interpolation*, the filling in of interior points between the sampled points, but as determined by the sampled points. The problem is made complex by the fact that smooth curves or sufaces which are tangent must be passed through many of the sampled points; it is not just a matter of connecting adjacent points with straight lines. Such an interpolation is referred to as *differentiable* or C^1. It is further complicated by the fact that the sampled data points may be irregularly scattered about on the surface or some two-dimensional projection of it.

Interpolation techniques for scattered data can be broadly classified into two categories: global and local. The *global* approach is exemplified by Hardy's (1970, 1990) multiquadric and Duchon's (1975) thin plate spline (TPS) methods. These techniques are effective and simple in their definition and, as noted by Franke (1982b), can produce visually pleasing surfaces. On the other hand, they involve the solution of large systems of linear equations that can become computationally explosive when dealing with even modestly large data sets. One remedy that has been sought to overcome this drawback is to localize these techniques over overlapping regions of the (x,y) domain (Franke, 1982b). Local methods involve the partitioning of the planar domain into nonoverlapping regions and defining the interpolants over the individual regions.

Triangle-based partitioning methods have received considerable attention due to some of their attractive features. Triangular regions introduce locality, numerical stability, and computational efficiency into the algorithms used to implement them. The study of piecewise interpolation techniques over triangulated domains has resulted in a wide variety of triangular interpolants, such as those reported by Barnhill et al. (1973); Farin (1985); Foley and Opitz (1992); and Nielson (1979). In most applications, in addition to obtaining a function that interpolates for given data, an additional constraint on the required global degree of smoothness is specified. Triangle-based methods that solve this problem, in addition to the function (z-axis) values at the vertices of the triangulation, also require estimates of the partial derivatives of the underlying function to be specified. In many situations the partial derivative information is not available and must be estimated from the f_i, x_i, y_i data itself.

8.1.1 Research Objectives

Any surface that fits the given function values (i.e., the f_i, z, or "depth" values of the surface at the two-dimensional domain sites) is only one of many possible solutions to the interpolation problem. Usually, the choice of the interpolant is governed by the degree of smoothness required, the ease of computation, and a subjective opinion of the visual aesthetics of the final surface. The final choice of the surface is determined as much by the application as the available interpolation tools. It can also be influenced by a priori knowledge of the origin of the scattered function values. All of these factors constrain the particular interpolated surface from the set of all possible surfaces.

In this chapter we discuss the analysis of C^1 interpolation for scattered data over triangulated domains. An additional constraint is introduced—that the final surface minimize a quantity that is expressed as a variational principle. The class of functionals considered are quadratics in the second-degree partial derivatives of the function. Since we are given only the function values at the domain sites to start with, it involves the estimation of the partial derivatives at the vertices of the triangulated domain. The minimizing principle in variational form is used as a constraint to estimate the unknowns at the vertices, i.e., the partial derivatives in the X and Y directions or the tangent planes. This is analogous to the work of Nielson (1979) and Pottmann (1992).

Techniques from the finite element method (FEM) of Strang and Fix (1973) is used to perform the minimization task. One class of FEM techniques involves the selection of *shape* or *basis* functions $\phi_1, \ldots, \phi_k$ associated with the unknowns $q_1, \ldots, q_k$ at the vertices of the elements (the domain triangles, in our case) and among all linear combinations $\Sigma_{i=1}^{k} q_i, \phi_i$, finding the one that minimizes the required functional. The unknowns q_i are then obtained as a solution to a system of equations (generally linear). In a scattered-data setting, the unknowns q_i are the partial derivatives at the vertices.

The accuracy and stability of the finite element method depends on the choice of the basis functions. In this part of our project, it was decided to use the hybrid cubic bézier triangular patch of Foley and Opitz (1992) as the basic building block to perform the minimization task. This triangular patch involves rational blending functions. Comparison studies are made with a polynomial triangular patch, the Clough–Tocher interpolant (Farin, 1985). These comparisons involve smoothness, accuracy, and numerical stability.

In the entire process described so far, there is one variable we haven't considered: the domain triangulation. The traditional approach in using triangular patches for scattered-data interpolation was to use an efficient and optimal algorithm to perform the triangulation of the domain sites. Initial work in most cases used the Delaunay triangulation (Lawson, 1977). Recent studies by Dyn et al. (1990) and Quak and Schumaker (1990) have shown that adapting the triangulation to reflect the given function values results in smoother surfaces. This aspect of adapting the triangulation based on the goodness of approximation of the FEM technique in minimizing the variational principle is studied.

Finally, extensions to using the proposed techniques on large data sets are presented. These are based on localizing the computational process over overlapping regions of the domain and performing the minimization process locally.

8.2 THE FINITE ELEMENT METHOD AND C^1 TRIANGULAR PATCHES

The finite element method has its roots in the fields of stress and strain analysis of structures. Engineers have been using FEM techniques to handle and solve the relatively complicated equations involved in elasticity and structural mechanics. The scope and applicability of the finite element method has grown rapidly to encompass a wide variety of engineering and scientific disciplines. However, we believe that this chapter may describe a very unusual application—to the topic of the visual perception of sampled surfaces. The literature in engineering and mechanics is copious. For general introductions to the basic principles and applications, see Burnett (1987) and Strang and Fix (1973).

In classical mathematical terms, the finite element method is an extension of the Rayleigh–Ritz–Galerkin technique as applied to a wide class of partial differential equations (Ciarlet, 1978). We are interested in the Ritz approximation technique (to be described shortly), which does not operate on the differential equation directly but on the equivalent variational form of the problem. Suppose the problem to be solved is given in variational form and it is required to find a given function f that minimizes a particular quantity. From classical variational calculus, we know that the minimization property leads to a differential equation (Fox, 1987) for f (the Euler equation). Normally, an exact solution to this problem is impossible and some approximation technique is necessary. The Ritz technique provides a stable and precise way of obtaining this approximation. The unknowns in the problem are generated as the solution to a system of equations (generally, a linear system) instead of to a differential equation. A brief description of the Ritz technique is now presented.

8.2.1 The Ritz Technique

The first step in obtaining the Ritz approximation (Strang and Fix, 1973) to a given problem is to partition the domain into a union of subdomains. Let us start by assuming that the domain has been partitioned into triangles using a particular algorithm, say the Delaunay triangulation technique.

The basic principle of the Ritz technique is to define basis functions ϕ_i associated with the parameters q_i at the vertices of the triangulation. The q_i are the function value, first partial in the X direction, first partial in the Y direction, and so on. At a vertex V_i the parameters associated with that vertex can be described as

$$q_i = D_i f(V_i) \tag{8.1}$$

where the differential operator D_i is of order zero if the parameter is the function value; it is $\partial/\partial x$ if the parameter is the partial derivative in X at that vertex; and so on. The basis functions ϕ_i are the cardinal basis functions associated with the parameter q_i. Therefore each ϕ_i has the property that $D_i\phi_i(V_i) = 1$. The cardinal basis functions therefore form a basis, since, over the triangle, every function f belonging to the space can be described as a linear combination $\Sigma_{i=1}^{n} q_i\phi_i$. Now, given a quadratic functional, e.g., the thin plate spline functional

$$I(f) = \iint f_{xx}^2 + 2f_{xy}^2 + f_{yy}^2 \tag{8.2}$$

substituting for f in the integral yields a quadratic in the parameters q_i at the vertices of the triangle. This can be solved to obtain optimum values of q_i.

This represents the minimization of the functional over a single triangle. Over a triangulated domain, the contributions to the quantity to be minimized are computed separately for each triangle. Noting that every vertex can be shared by more than one triangle, the contributions from all the triangles are *assembled* into one global system, which is a quadratic in the parameters at all the vertices of the triangulation.

The principle advantage of the technique is in the calculation of the quantity being minimized, locally over individual triangles, even though the final system that yields the unknowns is a global one. The success of the method depends heavily on the choice of the basis that bears on the accuracy and the stability of the technique.

8.2.2 C^1 Cubics over Triangles

The subject of interpolating for scattered data over triangulated domains has received considerable attention from the computer-automated graphics design (CAGD) community over the last two decades or so. The methods developed to date can be broadly classified into two categories, the *procedural* techniques and the *closed form* ones. Initial research in the field was oriented more toward the procedural type of methods. One class of triangle patch solutions involves the discretization of a transfinite interpolant, such as the BBG interpolant of Barnhill et al. (1973). This interpolant was developed based on the principle of the Boolean sum of three univariate interpolation operators. The side–vertex method of Nielson (1979) solves the problem in an analogous way. It consists of three univariate cubic interpolants in Hermite form along lines joining the vertices of the triangle and the edges opposite to them. The value of the interpolant at a point is obtained as a rational weighted combination of the three univariate interpolants. Nielson (1980) has developed a nine-parameter triangular interpolant based on a minimum-norm network over triangulated domains. An elegant technique falling into the class of closed-form triangular patches is the Clough–Tocher interpolant, which has its roots in classical FEM (Clough and Tocher, 1965). This was given a more geometrically oriented treatment and put into the Bernstein-Bézier form by Farin (1985). The Clough–Tocher method solves the problem by splitting each domain triangle into three microtriangles and then forming a piecewise cubic interpolant over the finer triangulation. Since it has a closed-form solution in the Bernstein Bézier form, this has the distinct advantage of possessing simple techniques to evaluate and also compute partial derivatives and surface normals required for visualization (Nielson et al., 1991) and geometry-processing applications (Mäntylä, 1988). The Clough–Tocher technique has been extended to higher dimensions by Worsey and Farin (1987).

Finally, we discuss the hybrid cubic Bézier triangular patch of Foley and Opitz (1992). The motivation for its development was to obtain a closed-form solution to some procedurally defined triangular patches, such as the side–vertex interpolant. This is accomplished via 12 control points instead of the 10 required for a standard Bézier patch. The control points on the boundary behave as in a standard Bézier triangle, but the inner point is obtained as a rational combination of three control points. The approach is similar to the one in Chiyokura and Kimura (1983), which represents a discrete version of Gregory's square with variable control points.

The Clough–Tocher triangular patch and the hybrid cubic Bézier triangular patch are of special interest to us and are described in more detail later.

The following sections describe the two triangular patches of interest to us: the cubic Clough–Tocher and the hybrid cubic Bézier triangular patches. We briefly discuss the function spaces over which these patches are defined. Before we do that, a brief note on the representation of cubic triangular Bézier patches is given.

8.2.3 Cubic Triangular Bézier Patches

Given 10 control points b_i where $|\mathbf{i}| = 3$ and a point with barycentric (i.e., measured from the center of mass) coordinates $\mathbf{u} = (u,v,w)$, the value of the cubic defined by the control net (Farin, 1993) evaluated at $\mathbf{u}$ is given as

$$b^3(\mathbf{u}) = \sum_{|i|=3} b_i B_i^3(u) \tag{8.3}$$

For nonparametric Bézier patches, we can assume that the domain points are in the XY plane and that the control points b_{ijk} have XY coordinates of the point whose barycentric coordinates are $\left(\frac{i}{3}, \frac{j}{3}, \frac{k}{3}\right)$. The value of the nonparametric patch at a point with barycentric coordinates $\mathbf{u} = (u,v,w)$ is now given analogously as

$$b^3(\mathbf{u}) = \sum_{|i|=3} b_i B_i^3(u) \tag{8.4}$$

8.2.4 The C^1 Clough–Tocher Triangular Patch

Suppose we are given a triangulation of the domain with associated function values and tangent planes. The Clough–Tocher triangular patch involves splitting of the domain triangle, called the *macro*triangle, into three small triangles called the *mini*triangles. This usually is done at the centroid of the triangle, as shown in Fig. 8.1. Three separate cubics are defined over the minitriangles (c,V_0,V_1), (c,V_1,V_2) and (c,V_2,V_1). For a detailed analysis of the C^1 Clough–Tocher interpolant, see Farin (1985).

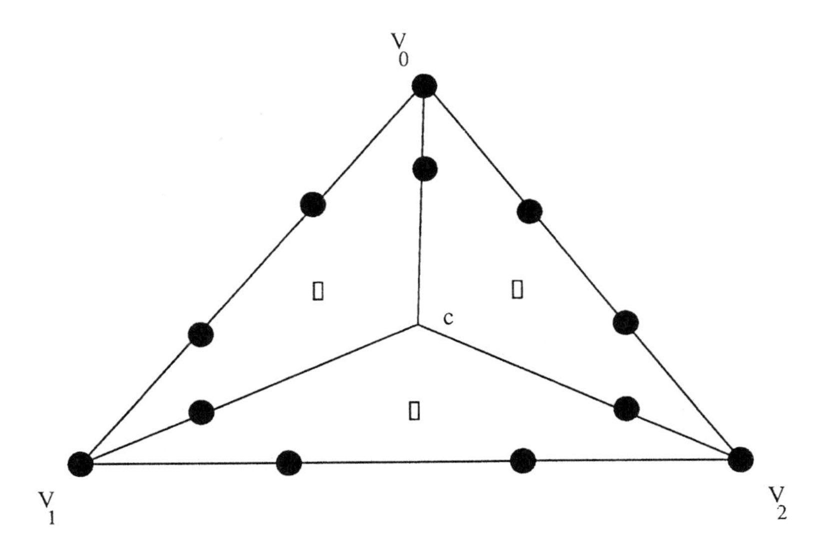

Figure 8.1 The Clough–Tocher Split. V_0, V_1, and V_2 are the vertices of the macrotriangle, and c is the location of the split point.

The Bézier ordinates of the boundary points can be computed by univariate cubic Hermite interpolation along the edges of the macrotriangle. These points are denoted by solid dots in Fig. 8.1. This leaves only the ordinate of the inner Bézier point to be determined. This is accomplished as follows: The inner point of each of the minicubics (shown by boxes in Fig. 8.1 is determined by considering the C^1 continuity conditions across adjacent macrotriangles. Once these points have been found, the points marked with crosses are computed from one C^1 continuity condition across adjacent minitriangles. The Bézier ordinate of the split point c is finally computed from the remaining C^1 condition involving the split point.

At this point we have all the information required to describe the piecewise cubic over the macrotriangle. The main step in the process, the computation of the Bézier ordinates of the inner points of the minitriangles is achieved as follows.

Consider two adjacent microtriangles sharing a common edge, as shown in Fig. 8.2. We can express c_1 and c_2 in terms of the barycentric coordinates of the opposite triangle as

$$c_1 = uc_2 + vV_1 + wV_0 \tag{8.5}$$

and

$$c_2 = uc_2 + vV_0 + wV_1 \tag{8.6}$$

The C^1 conditions across the edge require that the quadrilaterals formed by the points (b_1,a_2,c_1,a_1), (b_2,a_4,c_2,a_3) and (x,a_3,y,a_2) be coplanar. The first two conditions are satisfied by the construction of the boundary Bézier points. The third condition is enforced as follows: A direction $\mathbf{l}$ is chosen with its representation in barycentric coordinates with respect to the triangle (c_1,V_0,V_1) being (l_1,l_2,l_3). The directional derivative of the minicubic defined over triangle (c_1,V_0,V_1) along the direction $\mathbf{l}$ evaluated on edge V_0V_1 is a univariate quadratic polynomial whose Bézier ordinates are

$$3(l_1b_1 + l_2a_2 + l_3a_3), 3(l_1x + l_2a_3 + l_3a_2), 3(l_1b_2 + l_2a_4 + l_3a_3) \tag{8.7}$$

One technique used to compute the Bézier ordinates of the inner points of the microtriangles is to restrict this quadratic directional derivative to be linear; the condition can be expressed as

$$(l_1b_1 + l_2a_2 + l_3a_1) - 2(l_1x + l_2a_3 + l_3a_2) + (l_1b_2 + l_2a_4 + l_3a_3) = 0 \tag{8.8}$$

This can be solved for the unknown x, which is the Bézier ordinate of the inner point in the minicubic defined over (c_1,V_0,V_1). The unknown y in the microtriangle (c_2,V_1,V_0) can be found by an analogous procedure. It is essential that the direction $\mathbf{l}$ denote the same direction for both of the microtriangles. The usual

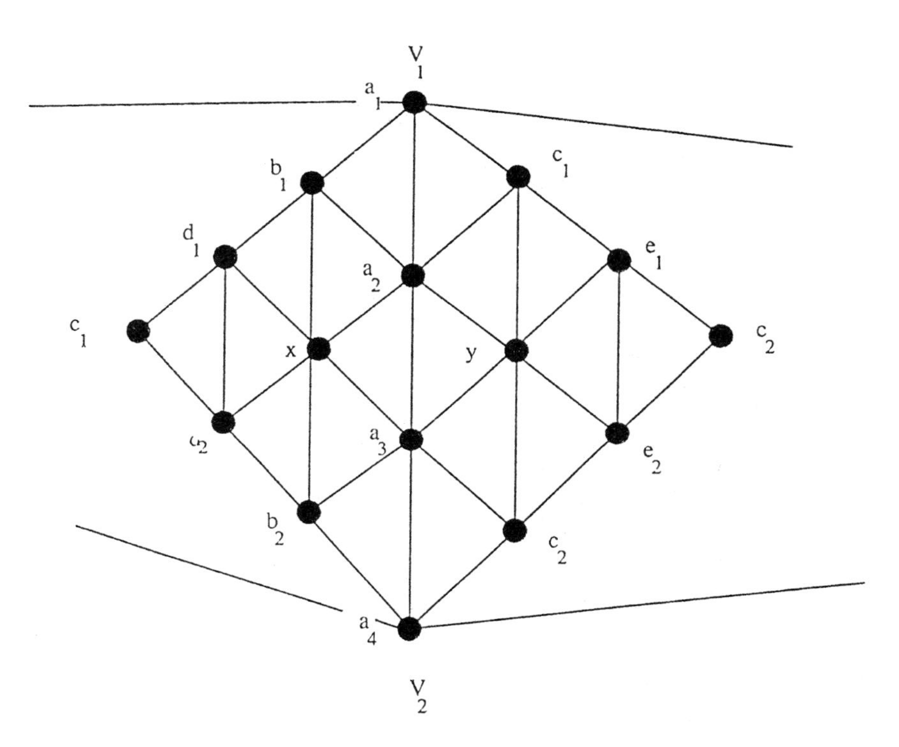

Figure 8.2 Adjacent C^1 patches over microtriangles. (c_1, V_0, V_1) and (c_2, V_1, V_0) are the vertices of the adjacent minitriangles, and c is the location of the split point.

approach taken to ensure this is to fix the direction $\mathbf{l}$ as being perpendicular to the edge V_0V_1.

The definition of the C^1 Clough–Tocher triangular patch with linearly varying cross-boundary derivatives is now complete. The C^1 cubic defined in this manner is found to be C^2 at the split point (Farin, 1985). Iterative methods to improve the quality of this interpolant have been investigated in Farin and Kashyap (1992). A technique to compute the inner points of the microtriangles based on minimizing the C^2 error across edges is presented in Farin (1985).

8.2.5 The Hybrid Cubic Bézier Triangular Patch

The definition of this triangular patch can be considered as a perturbation of the standard cubic Bézier triangular patch (Foley and Opitz, 1992). Suppose that we are given 12 control points p_0, p_1, p_2 and $b_i = b_{ijk}$, where $|\mathbf{i}| = 3$ and $|\mathbf{i}| \neq (1,1,1)$. The nonparametric hybrid patch is then defined by

$$H(\mathbf{u}) = \sum_{|I|=3} b_I B_I(\mathbf{u}), \tag{8.9}$$

where

$$b_{111} = b_{111}(\mathbf{u}) = w_0(\mathbf{u})p_o + w_1(\mathbf{u})p_1 = w_2(\mathbf{u})p_2 \tag{8.10}$$

and

$$\begin{aligned} w_0(\mathbf{u}) &= \frac{vw}{uv + uw + vw} \\ w_1(\mathbf{u}) &= \frac{uw}{uv + uw + vw} \\ w_2(\mathbf{u}) &= \frac{uv}{uv + uw + vw} \end{aligned} \tag{8.11}$$

The inner Bézier point is obtained (as can be seen from Foley and Opitz, 1992) as a rational weighted combination of the three points p_0, p_1, p_2. This construction allows a great deal of flexibility in controlling the cross-boundary derivatives across triangle edges. Each inner point p_i can be tied to a triangle edge (as is denoted in Fig. 8.3, where with an abuse of notation the p_i are shown to be located at different barycentric locations of the domain triangle). It can be seen that there is a singularity in the weight functions at the domain vertices. Even with these weight functions, the resulting patch is C^1 with removable singularities at the vertices. (For a proof, see Foley and Opitz, 1992.)

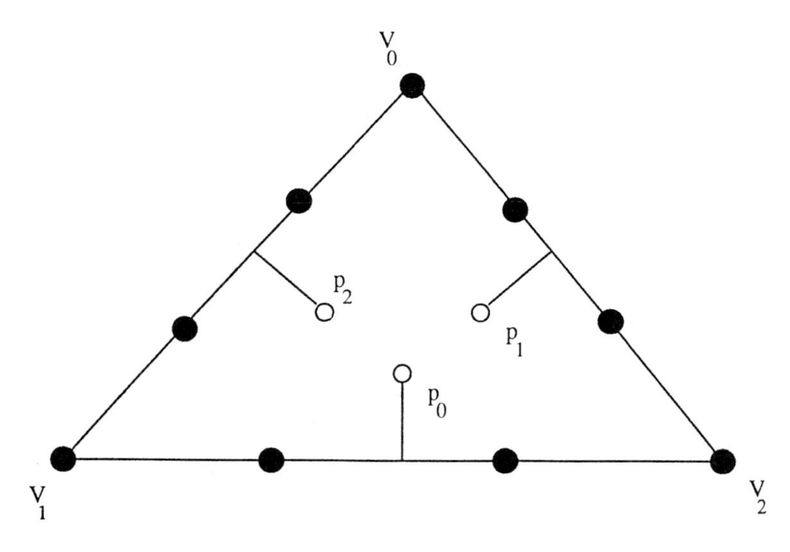

Figure 8.3 Symbolic representation of the hybrid cub Bézier triangular patch.

In the scattered-data setting, the Bézier ordinates of the boundary points are fixed by univariate cubic hermite interpolation along the edges of the triangle. The p_i can be computed in a procedure analogous to the computation of the inner point of microtriangles in the Clough–Tocher case (described earlier). The only difference is that although computing the point p_i associated with an edge V_jV_k of triangle (V_0,V_1,V_2), the direction $\mathbf{l}$ is defined with respect to triangle (V_0,V_1,V_2) itself, as opposed to the Clough–Tocher case, where it was defined over the microtriangle containing that edge.

Therefore with the linearly varying cross-boundary derivative assumption, all 12 of the Bézier ordinates needed to define the nonparametric hybrid cubic Bézier patch can be calculated.

We now return to the discussion of representing a function defined over a triangle with C^1 basis functions.

8.2.6 C^1 Functions Over Triangles

Given a triangle with vertices V_i, $i = 0, \ldots, 2$, with function values f_1, $\partial f_1/\partial x$, and $\partial f_1/\partial y$ the partial derivatives in X and Y, the function f over the triangle can be represented uniquely as

$$f(x,y) = \sum_{i=0}^{2} f_i b_i(x,y) + \frac{\partial f_i}{\partial x} c_i(x,y) + \frac{\partial f_i}{\partial y} d_i(x,y). \tag{8.12}$$

The b_i, c_i, and d_i are the appropriate cardinal basis functions with the following properties:

b_i:

$$\begin{aligned} b_i &= 1 \quad \text{at } V_i \text{ and 0 at other vertices} \\ \frac{\partial b_i}{\partial x} &= 0 \quad \text{at other vertices} \\ \frac{\partial b_i}{\partial y} &= 0 \quad \text{at other vertices} \end{aligned} \tag{8.13}$$

c_i:

$$\begin{aligned} c_i &= 0 \quad \text{at all vertices} \\ \frac{\partial c_i}{\partial x} &= 1 \quad \text{at } V_i \text{ and 0 at all vertices} \\ \frac{\partial c_i}{\partial y} &= 0 \quad \text{at all vertices} \end{aligned} \tag{8.14}$$

d_i:

$$
\begin{aligned}
d_i &= 0 \quad \text{at all vertices} \\
\frac{\partial d_i}{\partial x} &= 0 \quad \text{at all vertices} \\
\frac{\partial d_i}{\partial y} &= 1 \quad \text{at } V_i \text{ and 0 at all vertices}
\end{aligned}
\tag{8.15}
$$

Across triangle boundaries, the cross-boundary derivative is assumed to vary linearly. This is in agreement with the requirements for the basis functions described in Section 8.2.1.

8.3 FUNCTIONAL MINIMIZATION WITH C^1 TRIANGULAR PATCHES

In this section we present the theoretical and mathematical details of the minimization techniques for quadratic functionals with C^1 triangular patches. With basis functions defined in the Clough–Tocher and hybrid Bézier forms, the formulation of the minimization problem is described.

In a scattered-data interpolation setting, there have been several analogous efforts by researchers in the CAGD community. We list some of the salient efforts carried out in this field. The minimum norm network of Nielson (1983) was one of the earlier efforts. This involved the calculation of partial derivatives at the vertices of a triangulation by performing a minimization task on the curve network defined over the edges of the triangulation. This was extended to the case of functions defined on a spherical domain in Nielson and Ramaraj (1987). Pottmann (1991) presents a generalization of Nielson's technique to higher-order continuity. Pottmann (1992) has applied this principle to arbitrary three-dimensional domains. Renka and Cline (1984) presented their technique of determining derivative estimates based on minimizing a quadratic functional. They used basis functions defined radially around each data site in their approach. Alfeld (1985) has developed a minimization procedure for quintic Bézier polynomials using the q_{18} element of Barnhill and Farin (1973). Grandine (1987) solves a similar problem by performing a perturbation of the piecewise linear interpolant to the data to make it C^1.

We are interested in employing the hybrid cubic Bézier triangular patch as the basis for the minimization task. With its representation in the Bernstein Bézier form (de Boor, 1987), it can be thought of as the equivalent formulation of several procedurally defined triangular interpolants. For comparison studies, we use the Clough–Tocher triangular patch as a representative of the polynomial class of triangular interpolants.

We present the formulation of the minimization problem with the thin plate spline (TPS) functional as the quantity being minimized. We then generalize the technique to a wider class of quadratic functionals.

8.3.1 Minimization of the Thin Plate Spline Functional

The behavior of the TPS (Duchon, 1975) is related to the physical model of a simply supported, thin plate in a state of equilibrium. The surface of the plate in its equilibrium position can be represented by a function $f(x,y)$. This physical model can be approximated mathematically by invoking the following variational principle from classical mechanics:

> The deflection of a plate at equilibrium is that function f from a set V of admissible functions v for which the total strain energy $\varepsilon(f)$of the plate is minimal. (Courant and Hilbert, 1953)

In the form of an integral, the variational principle can be represented as

$$I(f) = \iint f_{xx}^2 + 2f_{xy}^2 + f_{yy}^2. \tag{8.16}$$

It is this variational form of the TPS that is employed in our minimization tasks. This problem can be solved for a given set of data points $(x_1,y_1), \ldots,$ (x_n,y_n) with associated function values. Subject to interpolation, the solution is of the form

$$Q(x,y) = \sum_{i=1}^{n} A_i d_i^2 \log d_i + a + bx + cy \tag{8.17}$$

where d_i is the distance between (x,y) and the point (x_i,y_i).

We are interested in performing this minimization according to the Ritz technique with the finite element method. In accordance with the principles of the technique, the problem is formulated one triangle at a time. The final global system of equations in all the unknowns is assembled from the linear systems for individual triangles. The details of the assembly of the global system are provided later. We start off by describing the problem over individual triangles.

Going back to our definition of a function over a triangle, given a triangle with vertices V_i, $i = 0, \ldots, 2$, with function values f_1, $\partial f_1/\partial x$, and $\partial f_1/\partial y$ the partial derivatives in X and Y, the function f over the triangle can be represented uniquely as

$$f(x,y) = \sum_{i=0}^{2} f_i b_i(x,y) + \frac{\partial f_i}{\partial x} c_i(x,y) + \frac{\partial f_i}{\partial y} d_i(x,y) \tag{8.18}$$

where b_i, c_i, and d_i are the cardinal basis functions. We now define a symmetric bilinear form over the space of C^1 cubic interpolants as

$$a(F,G) = \iint F_{xx}G_{xx} + 2F_{xy}G_{xy} + F_{yy}G_{yy} \tag{8.19}$$

Then the integral $I(f)$ in Eq. 8.16 can be written as

$$I(f) = \mathrm{a}(f,f) \tag{8.20}$$

Substituting Eq. 8.18 into the integral in Eq. 8.16, we can write the integral as

$$I(f) = \iint \left(\frac{\partial^2}{\partial x^2} \left(\sum_{i=0}^{2} f_i b_i(x,y) + \frac{\partial f_i}{\partial x} c_i(x,y) + \frac{\partial f_i}{\partial y} d_i(x,y) \right) \right)^2 +$$
$$\left(\frac{\partial^2}{\partial y^2} \left(\sum_{i=0}^{2} f_i b_i(x,y) + \frac{\partial f_i}{\partial x} c_i(x,y) + \frac{\partial f_i}{\partial y} d_i(x,y) \right) \right)^2 + \tag{8.21}$$
$$2\left(\frac{\partial^2}{\partial xy} \left(\sum_{i=0}^{2} f_i b_i(x,y) + \frac{\partial f_i}{\partial x} c_i(x,y) + \frac{\partial f_i}{\partial y} d_i(x,y) \right) \right)^2$$

We now introduce some notation. Let $\mathbf{q}$ denote the vector containing the partials in the X direction at the three vertices followed by the partials in Y and finally the function values. We can write $\mathbf{q}$ as

$$\mathbf{q} = \left[\frac{\partial f_0}{\partial x} \frac{\partial f_1}{\partial x} \frac{\partial f_2}{\partial x} \frac{\partial f_0}{\partial y} \frac{\partial f_1}{\partial y} \frac{\partial f_2}{\partial y} f_0 f_1 f_2 \right] \tag{8.22}$$

With the vector of the parameters at the triangle vertices denoted this way, we can write the quantity to be minimized as

$$\mathrm{a}(f,f) = \mathbf{q}^T \mathbf{K} \mathbf{q} \tag{8.23}$$

where the square matrix $\mathbf{K}$ contains integrals of the products of the second partials of the basis functions. Rather than describe this matrix explicitly, we perform some algebraic manipulations to reduce it into a more manageable form.

We can rewrite matrix $\mathbf{K}$ as

$$\mathbf{K} = \iint \begin{bmatrix} \frac{\partial^2 c_0}{\partial x^2} & \frac{\partial^2 c_0}{\partial y^2} & 2\frac{\partial^2 c_0}{\partial xy} \\ \frac{\partial^2 c_1}{\partial x^2} & \frac{\partial^2 c_1}{\partial y^2} & 2\frac{\partial^2 c_1}{\partial xy} \\ \vdots & \ddots & \vdots \\ \frac{\partial^2 b_3}{\partial x^2} & \frac{\partial^2 b_3}{\partial y^2} & 2\frac{\partial^2 b_3}{\partial xy} \end{bmatrix} \begin{bmatrix} 1 & 0 & 0 \\ 0 & 1 & 0 \\ 0 & 0 & \frac{1}{2} \end{bmatrix} \begin{bmatrix} \frac{\partial^2 c_0}{\partial x^2} & \frac{\partial^2 c_1}{\partial x^2} & \cdots & \frac{\partial^2 b_3}{\partial x^2} \\ \frac{\partial^2 c_0}{\partial y^2} & \frac{\partial^2 c_1}{\partial y^2} & \cdots & \frac{\partial^2 b_3}{\partial y^2} \\ 2\frac{\partial^2 c_0}{\partial xy} & 2\frac{\partial^2 c_1}{\partial xy} & \cdots & 2\frac{\partial^2 b_3}{\partial xy} \end{bmatrix} \tag{8.24}$$

On inspecting the matrix $\mathbf{K}$, we can see that its entries are of the form

$$K_{00} = a(c_0,c_0) \tag{8.25}$$

$$K_{01} = a(c_0,c_1) \tag{8.26}$$

and so on. Denoting the generic basis function by ϕ_l, the entries of **K** can be described as

$$K_{ij} = (\phi_i, \phi_j) \tag{8.27}$$

With one final manipulation, we are in a position to put the problem in a solvable form. Inspecting the vector **q**, we can see that the first six elements are the partials in X and Y at the vertices of the triangulation, i.e., the unknowns. Denoting this subvector as $\mathbf{q}_u$ and the basis functions associated with these unknowns (the c_i and the d_i) as σ_i, Eq. 8.23 can be written as

$$\mathrm{a}(f,f) = \mathbf{q}^T{}_u\mathbf{A}\mathbf{q} + \mathbf{b}\mathbf{q}_u + c \tag{8.28}$$

where

$$\begin{aligned} A_{ij} &= a(\sigma_i,\sigma_j) \\ b_i &= \sum_{j=0}^{2} a(\sigma_i,b_j)f_j + \sum_{j=0}^{2} a(b_j,\sigma_i)f_i \\ c &= a\left(\sum_{i=0}^{2} f_i b_i, \sum_{j=0}^{2} f_j b_j\right) \end{aligned} \tag{8.29}$$

It must be kept in mind that the σ_i are just an alternate representation for the cardinal basis functions associated with the partial derivatives. For example, the basis function σ_0 corresponds to c_0, σ_4 corresponds to d_1, and so on. They have been used only to simplify the representation of the entries of the matrix **A**. Also, for symmetric quadratic functionals such as the TPS functional, noting that

$$a(f,g) = a(g,f) \tag{8.30}$$

b can be simplified as

$$b_i = 2\sum_{j=0}^{2} a(\sigma_i,b_j)f_j \tag{8.31}$$

Setting the partials of Eq. 8.28 with respect to the vector of unknowns to zero we obtain the minimizing condition as

$$2\mathbf{A}\mathbf{q}_u + \mathbf{b} = 0 \tag{8.32}$$

The matrix **A** is of size 6×6, with **b** being a column vector of size 6.

With the basis functions being expressed in the Clough–Tocher or the hybrid Bézier form, the basis functions are available to us as functions of the barycentric coordinates of the triangle. In order to be able to compute the entries of **A** and **b**, we need to be able to differentiate them with respect to the global X and Y directions.

The minimization problem given by Eq. 8.32 is now in a soluble form. This can be solved to obtain the derivative estimates at the vertices of the triangulation. With the derivatives in hand we can use either the Clough–Tocher scheme of the hybrid Bézier method (or any other triangular interpolant) to reconstruct a C^1 surface to the scattered data.

8.3.2 Setting Up the global System

The minimization procedure described so far has been directed at the unknowns, i.e., the first partial derivatives, at the vertices of a single triangle. Noting that every vertex in our triangulation of the given data sites can be shared by more than one triangle, we need to combine the contributions to the quantity being minimized globally from all individual triangles.

This is one of the inherent advantages of the finite element method. Given a functional to be minimized, we find the entries of the coefficient matrix by computing the contributions to its elements from individual triangles. This is made possible by the fact that the cardinal basis functions for the parameters at a domain vertex (associated with the function value or the partial derivatives) are defined in piecewise fashion over their regions of support. Figure 8.4 shows the region of support of a basis function associated with a parameter value at a vertex. All the entries in the coefficient matrix containing this basis function is a cumulative sum of the contributions to that value from all the triangles containing the vertex under consideration.

Let $\mathbf{A}_i$ denote the coefficient matrix for triangle T_i with the unknowns at the vertices of triangle T_i being denoted by the vector $\mathbf{q}_i$. If $\mathbf{A}$ is the coefficient matrix of the global system, with $\mathbf{q}$ denoting the vector of all the unknowns, we can write

$$\mathbf{q}^T\mathbf{A}\mathbf{q} = \sum_i \mathbf{q}_i^T\mathbf{A}_i\mathbf{q}_i \tag{8.33}$$

with an analogous summation holding good for the vector $\mathbf{b}$ too.

This turns out to be a very simple operation. Consider the mth triangle in the triangulation with vertices (V_i, V_j, V_k). Let $\mathbf{q}_u^m$ be the vector of its unknowns and $\mathbf{A}_m$ be the matrix of coefficients associated with the solution vector. If $\mathbf{q}_u$ and $\mathbf{A}$ are the corresponding global vector of unknowns and coefficient matrix, respectively, we can easily determine the contribution of triangle m to the global system.

For $i = 1, \ldots, 6$, $\mathbf{A}[V_i][V_j] = \mathbf{A}[V_1][V_j] + A_i(ij)$. This is repeated for all triangles in the system. An analogous procedure is performed for computing the global $\mathbf{b}$ vector also. This is the final system that is solved to obtain the values of the unknowns at the vertices of the triangulation.

8.3.3 The Solution Method

For symmetric quadratic functionals, it can be proved that the resulting linear system is sparse, symmetric, and positive definite. This is guaranteed to have a

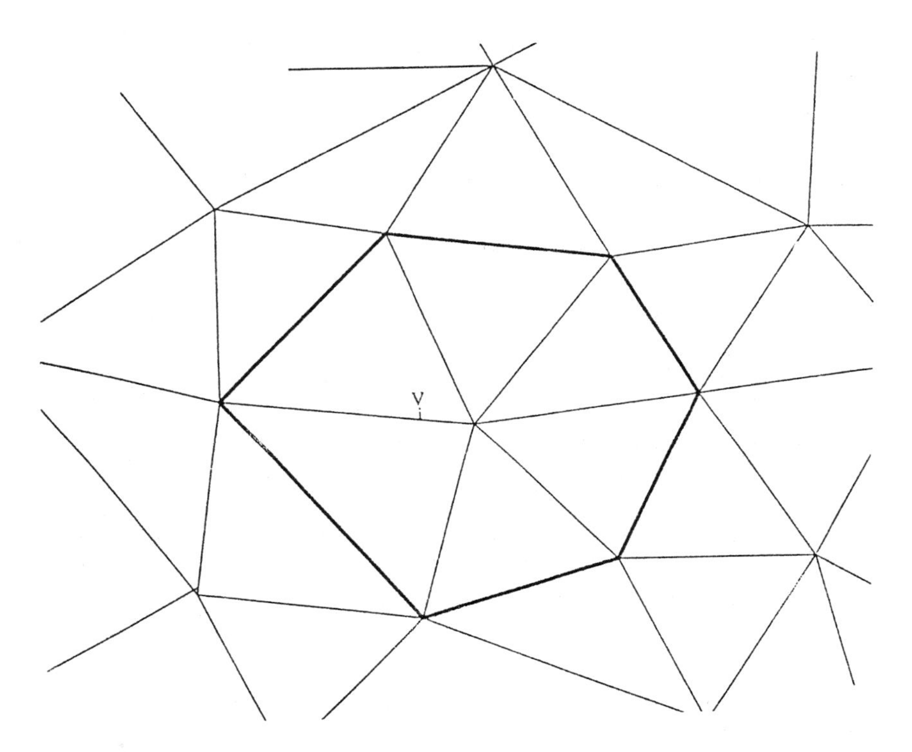

Figure 8.4 The region of support for a basis function associated with a vertex.

unique solution. The method used to solve the linear system has received considerable attention from the FEM community as a whole (Strang and Fix, 1973). The argument of direct versus iterative solutions prevails even today.

We employed a direct solution method using the Gaussian elimination technique with implicit scaled partial pivoting (Golub and van Loan, 1983).

8.3.4 Other Quadratic Functionals

The minimization technique described so far in this section was presented with the TPS functional as the variational principle to be minimized. The technique is a general one, and we now make some comments about the class of functionals to which it can be applied and provide some techniques for extending the process to them.

We restrict ourselves to only quadratic functionals in the second partial derivatives of the underlying functional. In the case of the TPS functional, the measure to be minimized globally over the surface was given in the form of an integral as

$$I(f) = \iint f_{xx}^2 + 2f_{xy}^2 + f_{yy}^2 \tag{8.34}$$

We then proceeded to define a symmetric bilinear form on the space of C^1 cubic interpolants as

$$a(F,G) = \iint F_{xx}G_{xx} + 2F_{xy}G_{xy} + F_{yy}G_{yy} \tag{8.35}$$

The key point of interest here is that the bilinear form so described be symmetric; i.e.,

$$a(F,G) = a(G,F) \tag{8.36}$$

This ensures that the coefficient matrix **A** in the solution to the problem is symmetric and positive definite. This guarantees the existence of a unique solution.

The minimization technique in the form described so far cannot be applied to functionals where $a(F,G)$ is not equal to $a(G,F)$. An example of such a functional is the square Laplacian that is given by

$$I(f) = \iint \nabla^2 f = \iint f_{xx}^2 + 2f_{xx}f_{yy} + f_{yy}^2 \tag{8.37}$$

The following subsection provides a way of generalizing the technique to arbitrary symmetric quadratic functionals.

8.3.5 Minimization of General Symmetric Quadratic Functionals

Since we are dealing with functionals involving the second partial derivatives in X and Y, let us denote the general form of such functionals as

$$I(f) = \iint \alpha f_{xx}^2 + \beta f_{yy}^2 + \gamma f_{xy}^2 \tag{8.38}$$

where α, β, and γ are scalar values. For example, in the case of the TPS functional, $\alpha = 1$, $\beta = 1$, and $\gamma = 2$. Functionals that can be described this way give us symmetric bilinear forms of the type $a(F,G) = a(G,F)$.

We can follow the same principle steps we employed for developing the solution to the minimization problem for the TPS for this general class too. Let the function f be represented in terms of the first and second partial derivatives as before. With **q** being the vector of parameters at the vertices of the triangulation, as before, we write f in a more general form as

$$f = \sum_{i=1}^{9} q_i \phi_i, \tag{8.39}$$

where the ϕ_i are the basis functions. Substituting this into Eq. 8.38 and putting it in matrix form we get

$$a(f,f) = \mathbf{q}^T\mathbf{K}\mathbf{q} \tag{8.40}$$

where the matrix **K** now becomes

$$\mathbf{K} = \iint \begin{bmatrix} \sqrt{\alpha}\dfrac{\partial^2\phi_1}{\partial x^2} & \sqrt{\beta}\dfrac{\partial^2\phi_1}{\partial y^2} & \sqrt{\gamma}\dfrac{\partial^2\phi_1}{\partial xy} \\ \sqrt{\alpha}\dfrac{\partial^2\phi_2}{\partial x^2} & \sqrt{\beta}\dfrac{\partial^2\phi_2}{\partial y^2} & \sqrt{\gamma}\dfrac{\partial^2\phi_2}{\partial xy} \\ \vdots & \ddots & \vdots \\ \sqrt{\alpha}\dfrac{\partial^2\phi_9}{\partial x^2} & \sqrt{\beta}\dfrac{\partial^2\phi_9}{\partial y^2} & \sqrt{\gamma}\dfrac{\partial^2\phi_9}{\partial xy} \end{bmatrix} \begin{bmatrix} \sqrt{\alpha}\dfrac{\partial^2\phi_1}{\partial x^2} & \sqrt{\alpha}\dfrac{\partial^2\phi_2}{\partial x^2} & \cdots & \sqrt{\alpha}\dfrac{\partial^2\phi_9}{\partial x^2} \\ \sqrt{\beta}\dfrac{\partial^2\phi_1}{\partial y^2} & \sqrt{\beta}\dfrac{\partial^2\phi_2}{\partial y^2} & \cdots & \sqrt{\beta}\dfrac{\partial^2\phi_9}{\partial y^2} \\ \sqrt{\gamma}\dfrac{\partial^2\phi_1}{\partial xy} & \sqrt{\gamma}\dfrac{\partial^2\phi_2}{\partial xy} & \cdots & \sqrt{\gamma}\dfrac{\partial^2\phi_9}{\partial xy} \end{bmatrix} \tag{8.41}$$

From this point on, the rest of the derivation for the coefficient matrix **A** and the vector **b** is as in the case of the TPS functional.

8.4 ASPECTS OF THE DOMAIN TRIANGULATION

In our discussion up to this point, one aspect of the solution to the problem has been implicitly assumed to be solved: the triangulation of the domain. Having made the choice of triangular patches or finite elements for minimizing problems in variational form, we haven't addressed this aspect in detail so far.

Ever since the study of triangular patches in the context of CAGD began, the triangulation of the domain and efficient algorithms to perform the same have received considerable attention. Given that there is no unique triangulation to a given set of points in the plane (Prenter, 1985; Lawson, 1986), researchers have attempted to design computationally efficient algorithms for this seemingly combinatorially inefficient task. Efforts to obtain a triangulation to a given point set, independent of the lexicographic ordering of points, by imposing a constraint on the nature of the triangulations formed have resulted in a variety of such criteria and hence triangulations.

For a long time, the Delaunay triangulation has been the triangulation method of choice in the CAGD community. It is very conveniently definable in terms of its geometric "dual," the Voronoi (1908) tesselation (also known as the Dirichlet, Thiessen, or Wigner–Seitz tesselation). This had been well known to mathematicians for a long time. It had been used in solving problems in geometry, geology, molecular physics and biology, and pattern recognition, among many other problem areas. It has been used for interpolating surfaces to scattered data by Farin (1990) and Sibson (1981). Optimal $O(n \log n)$-order algorithms (Aho et al., 1974) for obtaining the Delaunay triangulation of a point set have been developed. The triangulation also has the Min-Max property of maximizing the minimum angle of all the triangles in the final triangulation globally (Lawson, 1986). What makes it an attractive technique is that this minimization process can be performed locally, triangle by triangle, still ending up with a globally optimal solution. Nielson (1983) presents an affine-invariant norm to construct triangulations that are invariant to scale transformations.

In a scattered-data interpolation setting with piecewise triangular patches, the Delaunay triangulation has been used extensively. Almost all the developed methods relied on it, if for no other reason than to establish a standard for testing and comparing the interpolation technique with other triangular interpolants. It is only recently that the idea of data-dependent triangulations has received attention from researchers. The philosophy behind this class of triangulation techniques is to incorporate the function value available at triangle vertices as a factor in the triangulation, instead of just their spatial location in the domain plane.

The pioneering effort in this direction was by Dyn et al. (1990). They suggested that the quality of the piecewise linear interpolant to the function values at the vertices of a triangulation is a good indicator of the quality of the triangulation itself. They also suggested a number of criteria for evaluating the quality of the piecewise linear interpolant. Local procedures for modifying the triangulation based on these criteria were presented.

Quak and Schumaker (1989) present a technique for computing the "energy" of a triangular patch in terms of the Bézier ordinates of its control polygon. They use the familiar quadratic functional of the TPS as a measure of the energy of a triangular patch. In a later result (Quak and Schumaker, 1990), this measure is used to modify the triangulation using a local optimal method similar to the one presented in Dyn et al.'s (1990) report. This approach has some problems associated with it, since a fixed triangulation (the Delaunay) is used to compute derivative values at the vertices of the triangulation. These derivative values are then used in all further triangulation modification steps. It appears that this concept of tying the derivative estimation to a particular triangulation and then using these very estimates to modify it is not a very sound one.

In this section we present some efforts to modify the triangulation based on a "goodness" measure tied to the minimization technique used to estimate the derivatives themselves. In this manner, it is hoped that the dependence of a fixed triangulation on the derivative estimation is removed.

We start by presenting some of the principles of the local optimization process developed in Dyn et al. (1990), since it is applicable to the triangulation techniques we present later.

8.4.1 Locally Optimal Data-Dependent Triangulation Techniques

As stated earlier, the approach of Dyn et al. (1990) is based on fixing a criterion to determine the quality of the piecewise linear interpolant to the data. Since the function value at the vertices of a triangle uniquely determine the piecewise linear interpolant, the only variable in this process is the triangulation itself.

Let T be a triangulation of the given data. Based on the goodness criteria, a cost function $c(T)$ is associated with the triangulation. A triangulation T' is called optimal if $c(T') \leq c(T)$ for every triangulation T of the given point set.

The local optimization technique involves minimizing the cost function by swapping edges of the triangulation. Only the diagonals of convex quadrilaterals formed by two adjacent triangles are valid choices of edges to swap, since they do not affect the integrity of the triangulation. If e denotes such a valid choice of an edge to swap, the concept of an optimal edge is introduced.

An edge e is called locally optimal if either of the following is true: (a) the quadrilateral whose diagonal is e is not strictly convex, or (b) the quadrilateral is convex and $c(T) \leq c(T')$, where $c(T)$ and $c(T')$ are, respectively, the triangulations before and after the edge e is swapped.

Finally, a triangulation (T) is called a locally optimal triangulation if all the edges in the triangulation are locally optimal in the sense just described.

With these assumptions in hand, along with a reasonable cost function, we can implement relatively simple algorithms to obtain a locally optimal triangulation starting with an arbitrary one.

Dyn and his colleagues suggest two possible choices for the criteria used to define the cost functions associated with a triangulation. The first is called the jump-in-normal derivative (JND). This measures the jump in the normal to the piecewise linear interpolant across edges of the triangulation. The second, called the angle between normals (ABN), measures the actual angle between adjacent triangles forming an edge. The rationale behind these criteria is to obtain a good handle on how close the C^0 piecewise linear interpolant is to a C^1 surface. These criteria provide good results when applied to many standard functions and data sets.

8.4.2 Data- and Functional-Dependent Triangulation

The techniques just presented attempt to modify the triangulation based on the choice of the space of functions used in the minimization process for symmetric quadratic functionals. The local optimization procedure employed does not make use of the function values that are given at the vertices of the triangulation. In this section, we present a technique to tie the triangulation of the domain to the minimization of quadratic functionals and also to the function values at the domain data sites.

Quak and Schumaker (1990) present a method for modifying an initial triangulation based on an energy measure for all the triangles in the domain. Their method has the drawback that they use precomputed derivative estimates to calculate the energy at each stage of the local optimization process. We felt that this dependence on one set of derivative values to compute the energy is one that should be avoided.

The method we propose uses the minimization of the same quadratic functional that was used in Quak and Shoemaker (1990), the TPS functional, to compute the energy at each stage of the edge-swap test. The method differs in approach, in that the derivative estimates are computed again once the triangulation has been modified due to an edge swap. At each stage, the "energy" of the surface is evaluated using Eq. 8.28 which is,

$$a(f,f) = \mathbf{q}_u^T \mathbf{A} \mathbf{q}_u + \mathbf{b} \mathbf{q}_u + c \tag{8.42}$$

where the components in this equation are as explained in Section 8.3.1.

This quantity is evaluated to compute the energy, which is now the cost function of our triangulation. Edge swaps are performed as before in a local optimization procedure. The results of tests of this technique for both the Clough–Tocher and hybrid Bézier basis functions are presented in Section 8.6.

8.5 EXTENSIONS TO LARGE DATA SETS

The process of determining the partial derivatives in X and Y at the vertices of the domain triangulation described in the preceding sections is a global one. Noting that there are two unknowns per vertex, the size of the linear system to be solved is $2n \times 2n$, where n is the number of vertices in the triangulation. In the case where we utilize the extra degree of freedom associated with each edge in the triangulation, we have to solve a much larger linear system. A negative aspect of this is that for large data sets, the resulting linear system of equations can be ill conditioned and costly to solve. The choice of the basis functions also introduces another factor into the stability of the whole system, as described earlier.

In this section we present some techniques to localize the global process of minimizing the functional under consideration. The approach used is similar to the one presented by Franke (1982b). This is based on the concept of partitioning the domain into overlapping rectangular regions, forming local interpolants to the data in each region, and blending the local interpolants in a manner so as to yield a C^1 function.

The techniques presented here are all based on triangular partitioning of the domain. The weight functions are defined over the triangular regions in piecewise C^1 fashion. The general form of the technique is first described, followed by efficient ways of defining the weight functions.

The technique indicates that the quality of the final interpolant is tied directly to the quality of the local interpolants. The number and spatial position of the points used to form the local interpolants must therefore play an important role in determining the quality of the local interpolants. With this in mind, we also present some studies into efficient domain-partitioning techniques that depend on the spatial distribution of the data sites.

8.5.1 General Form of Localized Techniques

Suppose we are given n data sites (x_i,y_i) in the domain with associated function values f_i to which we need to interpolate a smooth surface. The first step in the process is to define the local regions S_i, $i = 1, \ldots, K$, in the domain, K being the number of regions. The union of all S_i should be the domain of interest. Franke (1982) uses rectangular partitions of the plane as a choice of regions. In general, the S_i should be bounded regions of the plane and not be mutually disjoint.

The necessary condition on the weight functions $W_j(x,y)$ for the combination of local interpolants to be a meaningful one is

$$\sum_{j=1}^{k} W_j(x,y) = 1 \tag{8.43}$$

for all (x,y) in the domain. Also, outside of S_j, $W_j(x,y) = 0$. The local interpolants in each region S_J are denoted by $Q_j(x,y)$. This is the interpolant to only the data sites contained in S_j. More precisely, $Q_j(x_i,y_i) = f_i$ for all $(x_i,y_i) \in S_j$.

Finally, the *localized interpolant* is defined as

$$F(x,y) = \sum_{j=1}^{K} W_j(x,y)Q_j(x,y) \tag{8.44}$$

Since the weight functions form a partition of unity, it is easy to see that $F(x,y) = f_i$ for $i = 1, \ldots, n$. The continuity class of the localized interpolant is the minimum of the continuity classes of the weight functions and the local interpolants; i.e., if $W_j \in C^\alpha$ and $Q_j \in C^\beta$, then $F \in C^{\min(\alpha,\beta)}$.

Franke (1982b) chooses the local regions by overlaying a rectangular grid over the domain and defining the weight functions as standard C^1 piecewise bicubic hermite patches. The local interpolants he uses are the TPS functionals. The problem with this approach is that if the data are dense in some areas of the domain and sparse in others, it is possible to end up with local regions with very few (and in extreme cases, no) data sites at all. To overcome this potential problem caused by the tensor-product grid structure, we have used regions defined as the union of triangles.

8.5.2 Selection of Local Regions

The objective of using triangular regions as opposed to the simpler case of rectangular ones is to obtain, in some sense, a more representative partitioning of the domain. The problems we can face with a simple rectangular grid of having an uneven distribution of points in each region (and, in extreme cases, no points) can be reduced due to the flexibility available with triangular regions. However, even with triangles, the distribution of points within regions can affect the quality of the local interpolants.

With these points in mind, an attempt was made to define regions for local interpolants based on the number of points contained in each region. The idea therefore is to introduce a fixed number of region vertices into the domain and to find optimal positions for these vertices, constrained by the requirement that in the final triangles, the approximate number of points per triangle is constant.

This leads to a constrained optimization problem where the unknowns are the positions of the vertices of the regions and the constraint is the membership per region. The time complexity of this problem increases with the number of data sites and the number of local regions required.

We opted to solve this task at hand by making use of the principle of *simulated annealing* (Kirkpatrick et al., 1982). This has been applied successfully in many applications in science and engineering where the solution space of the problem is too large to be exhaustively searched. Simulated annealing has been applied to the problem of obtaining data-dependent triangulations by a global optimization procedure by Schumaker (1993).

The concept and the applicability of simulated annealing is a well-known and -researched problem; therefore we present a very brief description of the technique as applicable to our problem.

Simulated Annealing and Domain Partitioning

Simulated annealing has its roots in the principles of thermodynamics that deal with the manner in which metals cool and anneal. At higher temperatures, atoms of the metal move freely with respect to each other. As the temperature drops, the mobility lessens. The metal in its fully cooled form can reach a minimum energy state with respect to the ordering of the atoms. In nature, it has been observed that slowly cooled metals tend to reach this minimum energy state more often than suddenly cooled or quenched metals.

The well-known distribution function from physics, the Boltzmann probability distribution, quantifies the behavior of a thermodynamical system in a particular state. This is given by the equation

$$\text{Prob}(E) \sim e^{(-E/kT)} \tag{8.45}$$

This equation expresses the idea that a system in thermal equilibrium at a temperature T has its energy probabilistically distributed among all different energy states E. Factor k is Boltzmann's constant. Therefore, it is possible for the thermodynamic system to go uphill as well as downhill from a particular point, but the probability of an uphill move is lower at lower temperatures.

It is precisely these principles on which the simulated annealing is based. Defining a measure of the energy of the minimization process at a particular stage, the probability that a system changes its energy from a value E_1 to E_2 is defined as

$$p = e^{[-(E_2-E_1)/kT]} \tag{8.46}$$

When $E_2 < E_1$, the move is always made. If the condition does not hold, then the move is still made if the value of p in Eq. 8.46 is more than that of a uniformly distributed random variable. This is often referred to as the Metropolis algorithm (Metroplis et al., 1953).

The important point to be noted is that the manner in which the system is *cooled*, called the *temperature schedule*, plays an important role in the success of the technique. At higher temperatures, the system is allowed to make riskier moves, with an understanding that this contributes overall to getting out of local minima. A more conservative approach is adopted at the final stages, i.e., at lower temperatures.

As applied to our problem, the required state of the vertices of the local regions is specified by the number of data sites contained in each region. The energy function is a measure of the deviation from the required number of points per region. The prescribed number of points are introduced at arbitrary locations and triangulated using the Delaunay technique. Then the annealing is started, with the spatial locations of the points being allowed to move around in the domain based on the Metropolis algorithm. At higher temperature, the points are permitted more mobility. The amount by which they move is progressively reduced as the temperature drops. This, in essence, prevents the system from making too drastic a mistake.

One example of such a partitioning is shown in Fig. 8.5. The data set consists of 100 data sites. The number of triangles in the domain partition is fixed at 14. This is done by fixing the convex hull of the partitions and thereby maintaining a constant triangle number. The final partitioning has approximately 7 points in each triangle.

Simulated annealing is not a cure-all for all constrained optimization tasks. The adaptability and applicability of the method depend solely on the problem at hand. Even after determining this, it takes a lot of practice to come up with a good annealing schedule for the problem. For the domain partitioning prob-

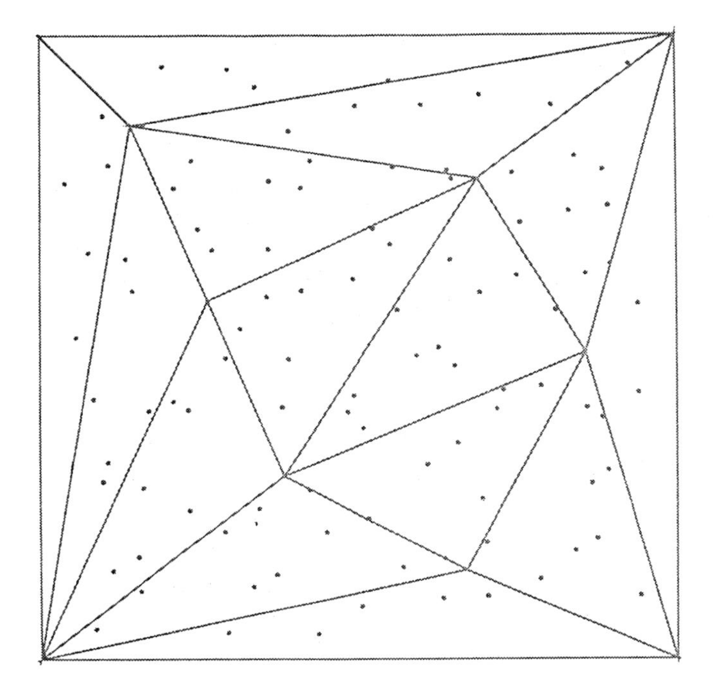

Figure 8.5 An example of a triangular domain partitioning using simulated annealing.

lem, with the number of data sites being in the hundreds, it was possible to fix a good annealing schedule, but only after some experimentation.

8.5.3 Localizing the Functional Minimization Problem

The entire motivation for the techniques presented up to this point in this section has been to find an alternative to solving the global $2n \times 2n$ system of equations for very large data sets. Using the principle of localizing the process over overlapping regions of the domain, we would like to obtain efficient and accurate ways of performing the minimization process locally.

With respect to the minimization problem, this brings to notice one immediate problem with the localized methods described in preceding sections. We started out by introducing a fixed number of points at certain locations in the domain (either arbitrarily or based on a predetermined criterion). If we attempt to carry out our minimization process in some way locally over these regions, we can see that the convex hull of the data sites in each region, in most cases, does not cover the entire region. Assuming that we obtain derivative estimates locally within regions, then forming the local interpolants is not possible at points that lie outside the convex hull of the data sites but within valid interior of a region. This situation is shown in Fig. 8.6.

We attempted to solve this problem via a modification of the simulated annealing technique used to obtain the regions. Instead of having the entire convex hull of the data sites as the set of possible locations for the vertices of the partitions, we restrict the location of the vertices to the actual data sites themselves. With this modification of the problem definition, the regions now obtained coincide exactly with the convex hulls of the data sites. The implementation of the Metropolis algorithm is, however, a bit more complex now, since we have to make sure that the vertices of the triangular regions always correspond to data sites. Even though the possible locations for the region vertices is a smaller set now, compared to the infinite possibilities existing in the earlier

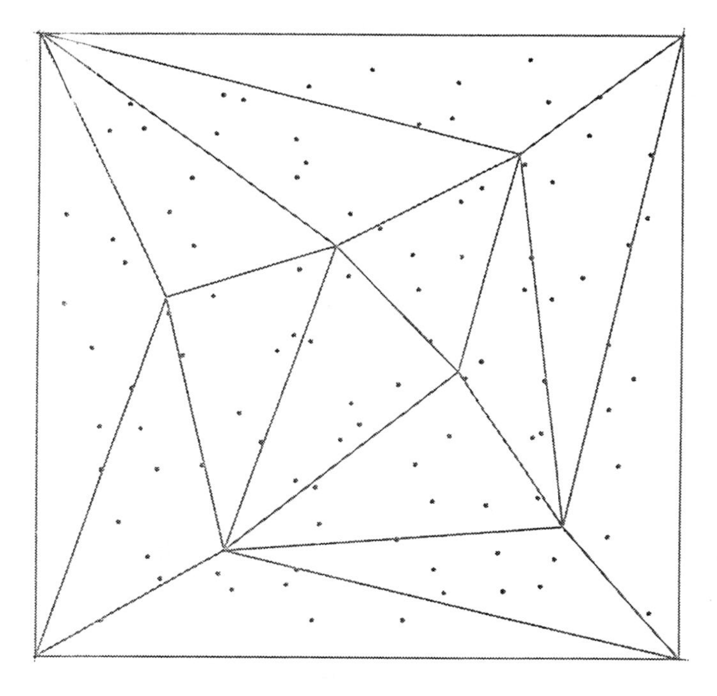

Figure 8.6 An example of a triangular domain partitioning where the convex hull of the data sites does not coincide with the region.

case, the additional work that must be done to select them increases the time complexity of the annealing process overall. This problem can be alleviated by means of a procedure that precomputes the nearest neighbors of every data site in a preprocessing step. This list of nearest neighbors can be used to advantage during the annealing procedure.

Assuming now that we have a valid triangular partitioning of the data sites, and weight functions defined over them, the localized interpolant at a point (x,y) is formed as follows. We first determine the partition triangle in which the point (x,y) lies. This triangle is common to exactly three regions. We then perform a minimization of the functional over a triangulation of the data sites in each region. This involves solving three linear systems, one corresponding to each of the three regions that contain the evaluation point. These linear systems need not be solved afresh for each evaluation point. Once the regions are fixed, a list of all the points in each region can be generated, and the FEM solution to the functional being minimized can be computed once and stored. This reduces the derivative estimation operation to just a lookup operation during the computation of the local interpolants.

We then form the local interpolant at the point for each region, based on the derivatives estimated locally over that region. Finally, we form the weighted sum of the local interpolants using the weight functions defined over the regions as before.

This method has the advantage that the solution to the global system is now replaced by a sequence of much smaller linear systems over each region. Additional considerations, such as having the local derivative estimation based on data-dependent triangulations with the same or different criteria in each region, can also be tried out.

The next section presents the results of localizing the minimization process for large data sets.

8.6 RESULTS

In this section we present the results of the techniques we have proposed so far in this chapter. The minimization for the class of symmetric quadratic functionals described in the preceding sections was tested out on a variety of standard test functions and data sets. We have used the data sets and functions prescribed by Franke (1987) for evaluating the performance of the proposed techniques.

The quality of the interpolants produced by the techniques are compared using the standard techniques of contours and isophote plots (Hagen et al., 1990). All the surfaces were evaluated on a square grid laid over the domain, which happens to span $[0,1] \times [0,1]$ for most of the test functions.

The implementation of all the proposed techniques was carried out on our laboratory's Silicon Graphics workstation. Surfaces were visualized by using a

Figure 8.7 A contour plot of Franke-1 with Clough–Tocher basis on 33-point set for the TPS functional with linear cross-boundary derivatives.

standard Phong lighting method with Gouraud shading (J. D. Foley et al., 1990). Surface plots of many of the results are also included in this section. We now present a discussion of the results.

8.6.1 Tests for Smoothness and Accuracy

The two basis functions—in the Clough–Tocher and Hybrid Bézier forms—were compared for smoothness and accuracy. A number of symmetric quadratic functionals were minimized using the two kinds of basis functions. Figure 8.7 shows the contour plot of the surface obtained by minimizing the TPS functional on the 33-point data set, with data sampled from the function Franke-1 and using the basis functions in Clough–Tocher form with linear cross-boundary derivatives. Figure 8.8 shows the isophote plot for the same surface.

The corresponding contour and isophote plots with the hybrid Bézier type of basis functions and linear cross-boundary derivatives is shown in Figs. 8.9 and 8.10. The performance of the two bases is of the same order; but on the whole,

Figure 8.8 An isophote plot of Franke-1 with Clough–Tocher basis on 33-point set for the TPS functional with linear cross-boundary derivatives.

Figure 8.9 A contour plot of Franke-1 with hybrid Bézier basis on 33-point set for the TPS functional with linear cross-boundary derivatives.

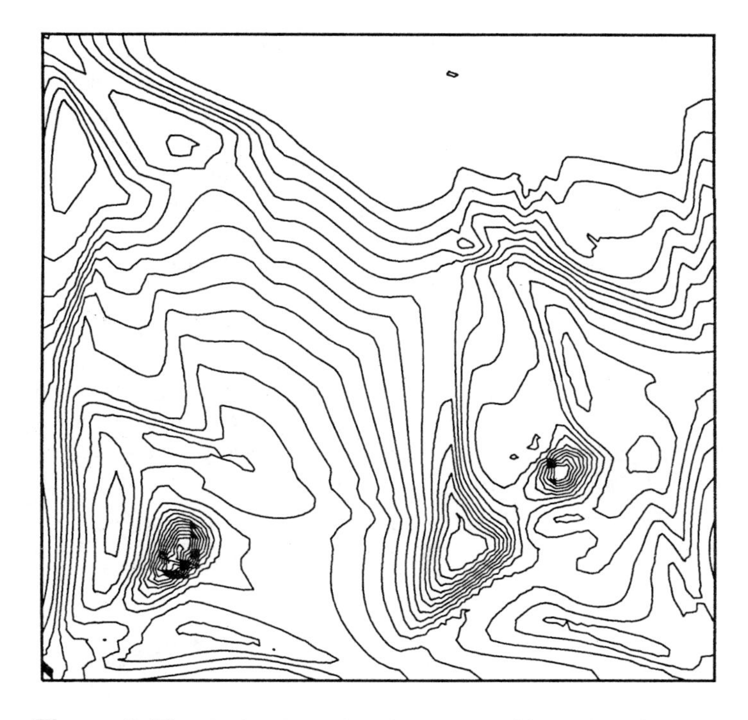

Figure 8.10 An isophote plot of Franke-1 with hybrid Bézier basis on 33-point set for the TPS functional with linear cross-boundary derivatives.

the Clough–Tocher basis tends to yield a smoother interpolant, as can be seen from the plots.

8.6.2 Stability Analysis

The linear systems that are solved in the minimization process for the unknowns were compared for their stability. The linear systems produced by both choices of basis functions have the same sparseness. We used the condition number of the linear system as an index into the stability of the systems.

It was found that over a wide range of triangulations and test functions, the Clough–Tocher basis functions produced a considerably more stable system. This is due predominantly to the fact that the Clough–Tocher is a polynomial triangular patch, as opposed to the rational blending functions used in the hybrid Bézier patch. Another possible factor that can influence the stability is the perturbation to the true solution introduced by the numerical integration performed in the case of the hybrid Bézier type of basis functions.

8.6.3 Data and Functional Dependent Triangulations

The process of carrying out local optimization for modifying the triangulation of the domain was tested out on different functions and data sets. The first criteria used was the length of the longest edge in each triangle. This was minimized for both choices of the basis functions. The new triangulation obtained by using this criterion is shown in Fig. 8.11. The contour and isophote plots for Franke-1 defined over the new, modified triangulation are shown in Figs. 8.12 and 8.13, respectively, for the Clough–Tocher basis functions. The corresponding plots for the hybrid Bézier basis functions are shown in Figs. 8.14 and 8.15. Some improvements in the quality of the interpolant can be seen in the case of the modified triangulation. This is observed for both choices of the basis functions. The minimization tends to perform better when we have triangles with smaller edges, as can be seen even though the improvements are not dramatic.

The second criterion used for the modification of the triangulation was the condition number of the global system. This was performed locally also, with

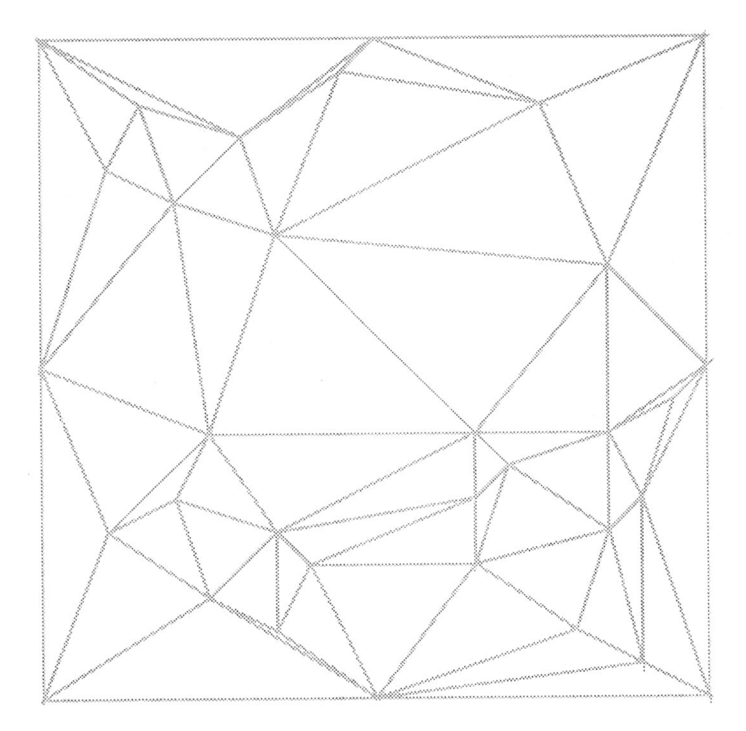

Figure 8.11 Triangulation of 33-point data set by minimizing the lengths of the longest edges.

Figure 8.12 A contour plot of function Franke-1 over triangulation, with minimizatino of longest edges and Cough–Tocher basis functions.

Figure 8.13 An isophote plot of function Franke-1 over triangulation, with minimization of longest edge and Clough–Tocher basis functions.

Figure 8.14 A contour plot of function Franke-1 over triangulation, with minimization of hybrid Bézier basis functions.

Figure 8.15 An isophote plot of function Franke-1 over triangulation, with minimization of longest edge and hybrid Bézier basis functions.

candidate edges being swapped when an improvement in the condition number was obtained. The triangulation obtained for the 33-point data set by locally optimizing this criterion with the Clough–Tocher basis functions is shown in Fig. 8.16. The isophote plot of the surface obtained is shown in Fig. 8.17. The original surface shows instabilities (seen as erratic or oscillatory behavior) in regions of the domain. The surface defined over the modified triangulation reduces this behavior. The more stable system more often leads to better surfaces, as can be expected. The corresponding isophote plot with the hybrid Bézier basis functions is shown in Fig. 8.18.

The third criterion used was a measure of the "energy" of the surface as given by the TPS functional. This was minimized using a local optimization process. The triangulation obtained for the 33-point data set is shown in Fig. 8.19. The next two figures show the isophote plots of the surface obtained by minimizing the TPS functional over the modified triangulation for the two

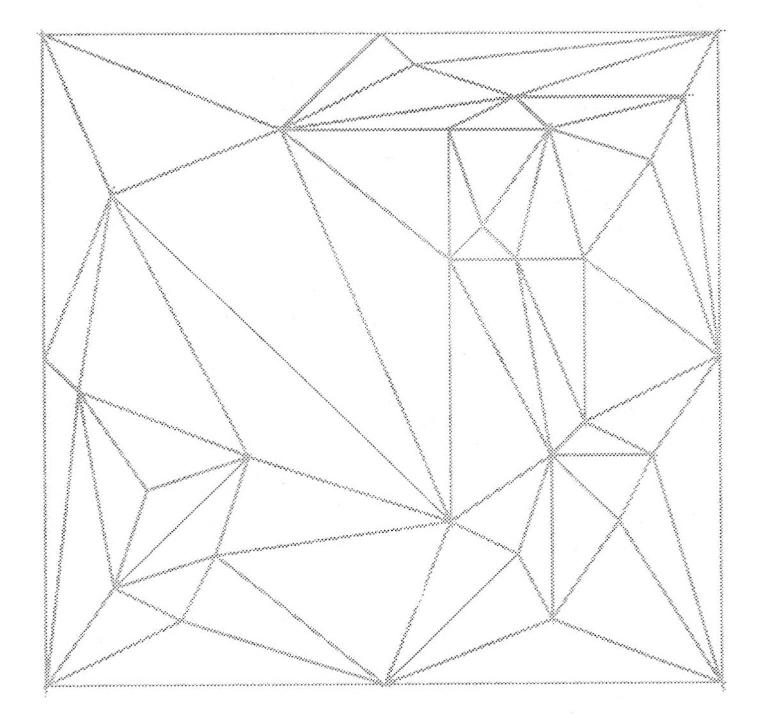

Figure 8.16 Triangulation of 33-point data set by minimizing the condition number of the global system.

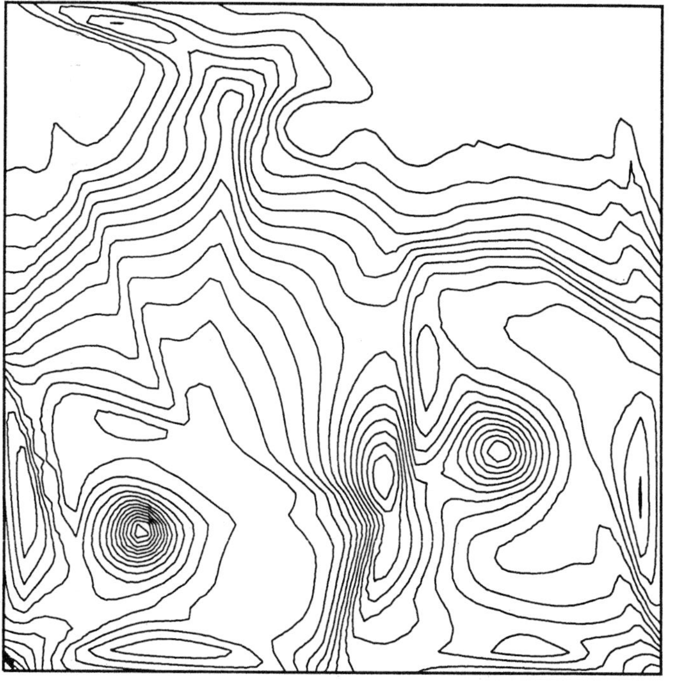

Figure 8.17 An isophote plot of function Franke-1 over triangulation, with minimization of the condition number with Clough–Tocher basis functions.

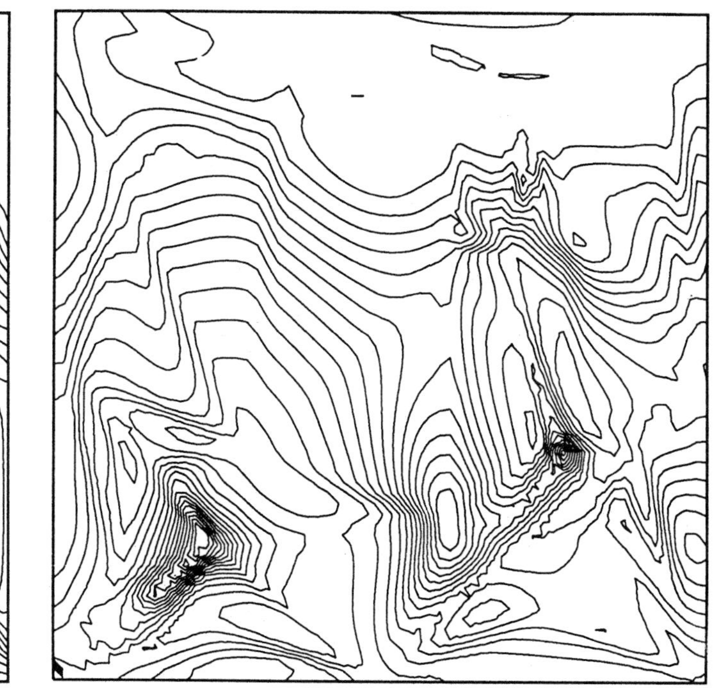

Figure 8.18 An isophote plot of function Franke-1 over triangulation, with minimization of the condition number with hybrid Bézier basis functions.

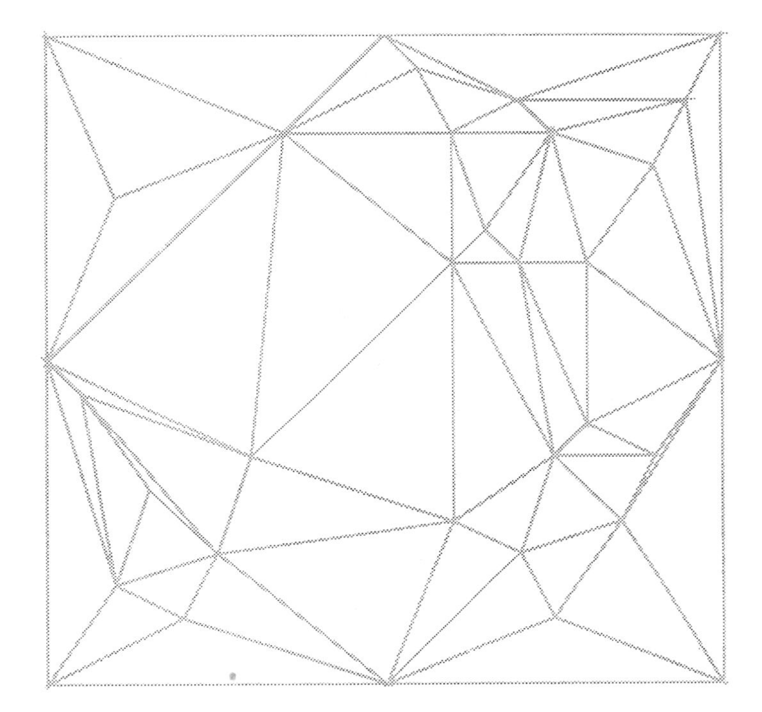

Figure 8.19 Triangulation of 33-point data set by minimizing the "energy" of the surface.

choices of function spaces. These surfaces can be seen to have significant improvements over the ones obtained with the other triangulation criteria.

8.6.4 Extensions to Large Data Sets

The localization of the global minimization process was carried out on a randomly generated 200-point data set. The regions were obtained by the process of simulated annealing. Figure 8.20 shows a randomly generated 200-point data set partitioned into 40 triangles using the annealing technique. Each triangle can be seen to have approximately 5 points. Figure 8.21 shows the contour plot of the surface obtained by performing the minimization process locally using the triangular regions overlaid on the 200-point data set and Clough–Tocher basis functions. Figure 8.22 shows the isophote plot for the localized approach for the same data set. Figures 8.23 and 8.24 show similar plots for the case of the hybrid Bézier basis functions.

We can see that the localized technique performs quite reasonably. The er-

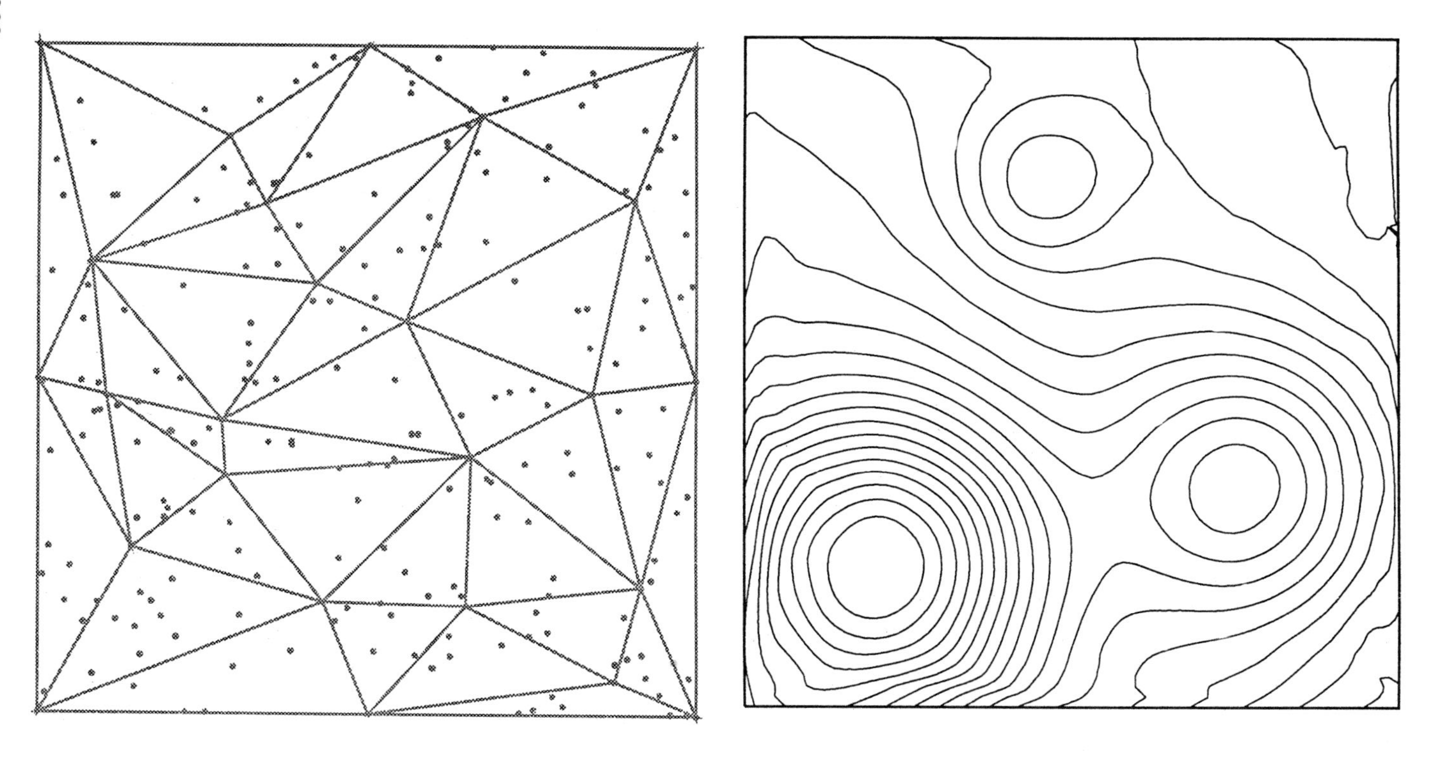

Figure 8.20 Domain partitioning for 200-point data set using simulated annealing.

Figure 8.21 A contour plot of Franke-1 using the localized method with Clough–Tocher basis functions.

Figure 8.22 An isophote plot of Franke-1 using the localized method with Clough–Tocher basis functions.

Figure 8.23 A contour plot of Franke-1 using the localized method with hybrid Bézier basis functions.

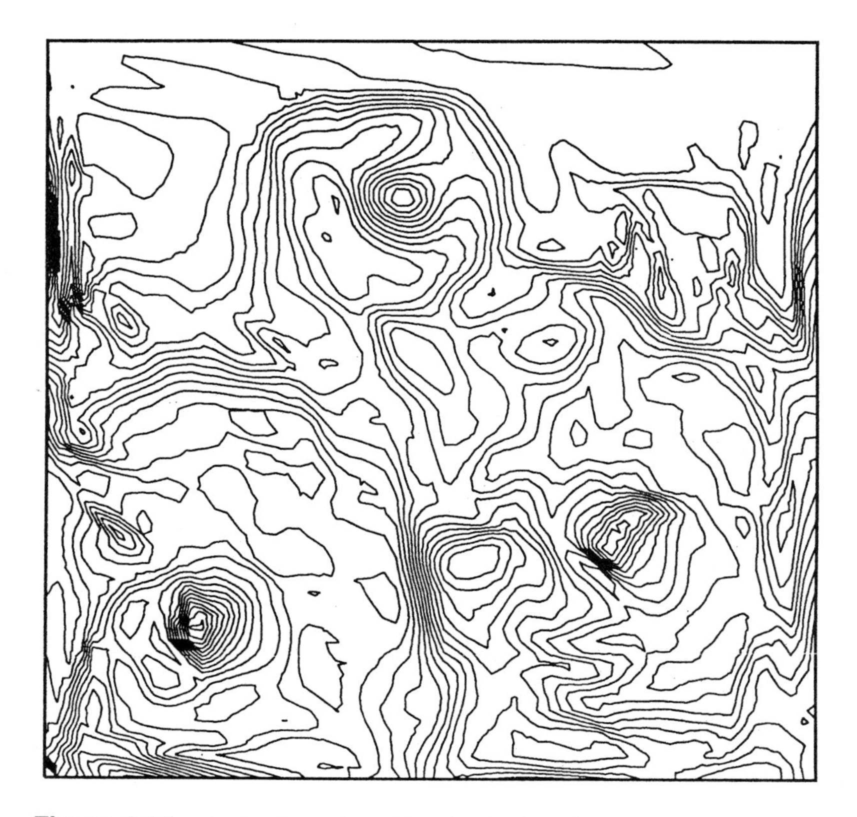

Figure 8.24 An isophote plot of Franke-1 using the localized method with hybrid Bézier basis functions.

ratic behavior present in regions of the surface coincide with the regions where even the global technique has problems too. Generally, the performance of the localized technique is of the same order as the global one. As in the other examples, the Clough–Tocher basis functions perform better than the hybrid Bézier ones. Since they are not as stable as the Clough–Tocher basis functions, the global and the localized techniques show more erratic behavior, as can be seen from the isophote plots.

With the linear systems solved being of a much lower order, this makes the localized method an attractive one for using it with large data sets. The performance can be considerably improved if a data reduction strategy is used to eliminate data sites wherever possible, thereby aiming to improve the stability of the localized approach.

8.7 CONCLUDING REMARKS

The main objective of this research was to perform the minimization of symmetric quadratic functionals over triangular domains. With a good understanding of the theoretical background of the problem, an attempt was made to develop efficient techniques and algorithms to perform the minimization task. Techniques from the finite element method were used to perform the minimization of a class of quadratic functionals, the symmetric ones.

We also decided to use a relatively new triangular patch, the hybrid cubic Bézier, for the minimization task. Recognizing that the hybrid cubic Bézier triangle contains rational terms, we also developed the technique to handle a counterpart from the polynomial class of triangular interpolants, the Clough–Tocher triangular patch. The smoothness and accuracy studies for the two function spaces for which the patches form a basis reveals that the polynomial space S^1_{CT} patch tends to perform better usually, when compared to S^1_{HB}, the rational one.

An attempt was also made to study the effects of the domain triangulation on the accuracy of the minimization process. Using criteria derived from error analysis studies for FEM problems, we designed criteria to associate cost functions with the triangulation of the domain. By tying the triangulation to the functional minimization task, it was possible to obtain smoother surfaces to scattered data sets. It was also observed that the Delaunay triangulation performs reasonably well in most cases, compared to the modified triangulation techniques. The trade-off in the improvements we can obtain with the data- and functional-dependent triangulations is the computational cost of obtaining the triangulations.

This method of reconstructing surface from sampled and scattered data points is offered as a plausible model of how the process of filling surfaces is handled psychophysically when stimuli are of this same general kind. In previous experimental work in this laboratory (Uttal et al., 1988) we have observed exactly this process to be taking place in human perceptual processing. This formal model is offered as a description of those psychophysical processes. Obviously, the same mathematical functions are unlikely to be carried out in the organic brain, but processes that are analogous to these formal methods must be instantiated in some form there.

NOTE

1. This chapter is adapted from a dissertation entitled "Data and Functional Dependent Interpolation over Triangulations" submitted by Sriram Dayanand (1993) in partial fulfillment of the requirements for the Ph.D. degree from the Department of Computer Science at Arizona State University.

Bibliography

Abidi, M. A., and Gonzales, R. C. (1992). *Data Fusion in Robotics and Machine Intelligence*. San Diego: Academic Press.

Adams, E. Q. (1923). A theory of color vision. *Psychological Review, 30*, 56–76.

Aho, A., Hopcroft, J., and Ullmann, J. (1974). *The Design and Analysis of Computer Algorithms*. Reading: Addison-Wesley.

Alfeld, P. (1985). Derivative generation from multivariate scattered data by functional minimization. *Computer Aided Geometric Design, 2*, 281–296.

Allman, J. (1981). Reconstructing the evolution of the brain in primates through the use of comparative neurophysiological and neuroanatomical data. In E. Armstrong and D. Falk (eds.), *Primate Brain Evolution: Methods and Concepts* (pp. 13–28). New York: Plenum.

Aloimonos, J., and Shulman, D. (1989). *Integration of Visual Modules*. San Diego: Academic Press.

Anderson, J. A. (1968). A memory model utilizing spatial correlation functions. *Kybernetik, 5*, 113–119.

Anderson, J. A., and Rosenfeld, R. E. (eds.). (1988). *Neurocomputing: Foundations of Research*. Cambridge, MA: MIT Press.

Anderson, J. A., Pellionisz, A., and Rosenfeld, E. (eds.). (1990). *Neurocomputing: Vol. 2. Directions for Research*. Cambridge, MA: MIT Press.

Argyle, E. (1971). Techniques for edge detection. *Proceedings of the IEEE, 59*, 285–287.

Attneave, F. (1950). Dimension of similarity. *American Journal of Psychology, 3*, 516–556.

Attneave, F., and Arnoult, M. D. (1956). The quantitative study of shape and pattern perception. *Psychological Bulletin, 53*, 452–471.

Barnhill, R. E. (1977). Representation and approximation of surfaces. In J. R. Rice (ed.), *Mathematical Software III* (pp. 69–120): Academic Press.

Barnhill, R. E. (1985). Surfaces in computer-aided geometric design: A survey with new results. *Computer-Aided Geometric Design*, *2*, 1–17.

Barnhill, R. E., and Farin, G. (1973). C1 quintic interpolation over triangles: Two explicit representations. *International Journal for Numerical Methods in Engineering*, *17*, 1763–1778.

Barnhill, R. E., Birkhoff, G., and Gordon, W. J. (1973). Smooth interpolation in triangles. *Journal of Approximation Theory*, *8*, 114–128.

Baruch, O., and Loew, M. H. (1988). Segmentation of two-dimensional boundaries using the chain code. *Pattern Recognition*, *21*, 581–589.

Beiderman, I. (1987). Recognition-by-components: A theory of human image understanding. *Psychological Review*, *94*, 115–147.

Beiderman, I., and Ju, G. (1988). Surface versus edge-based determinants of visual recognition. *Cognitive Psychology*, *20*, 38–64.

Bennett, B. M., Hoffman, D. D., and Prakash, C. (1989). *Observer mechanics: A formal theory of perception*. San Diego: Academic Press.

Bennett, B. M., Hoffman, D. D., and Murthy, P. (1993). Lebesgue logic for probabilistic reasoning and some applications to perception. *Journal of Mathematical Psychology*, *37*, 63–103.

Blake, A., Zisserman, A., and Knowles, G. (1989). Surface descriptions from stereo and shading. In B. K. P. Horn and M. J. Brooks (eds.), *Shape from Shading* (pp. 29–52). Cambridge, MA: MIT Press.

Bronshtein, I. N. and Semendyayev, K. A. (1973). A *Guide Book to Mathematics: Fundamental Formulas, Tables, Graphs, and Methods*. New York: Springer-Verlag.

Brown, J. L. (1965). The structure of the visual system. In *Vision and Visual Perception*, (Graham, C.H., ed.). New York: Wiley. (pp. 39–59).

Brown, P. K., and Wald, G. (1964). Visual pigments in single rods and cones of the human retina. *Science*, *144*, 42–52.

Bülthoff, H. H., and Mallot, H. A. (1990). Integration of stereo, shading, and texture. In A. Blake & T. Troscianko (eds.), *AI and the EYE* (pp. 119–146). New York: Wiley.

Bundesen, C., and Larsen, A. (1975). Visual transformation of size. *Journal of Experimental Psychology: Human Perception and Performance*, *1*, 214–220.

Burnett, D. S. (1987). *Finite Element Analysis, From Concepts to Applications*. Reading, MA: Addison-Wesley.

Burns, J. B., Hanson, A. R., and Riseman, E. M. (1986). Extracting straight lines. *IEEE Transactions on Pattern Analysis and Machine Intelligence*, *8*, 425–455.

Caelli, T., and Reye, D. (1993). Classification of images by color, texture, and shape. *Pattern Recognition*, *26*, 461–470.

Caelli, T., Johnstone, M., and Robison, T. (1993). 3-D object recognition: Inspirations and lessons from biological vision. In A. Jain and P. Flynn (eds.), *Three-Dimensional Object Recognition Systems* (pp. 1–17). New York: Elsevier.

Campbell, F. W. and Maffei, L. (1981). The influence of spatial frequency and contrast on the perception of moving patterns. *Vision Research*, *21*, 713–721.

Campbell, F. W., and Robson, J. G. (1968). An application of Fourier analysis to the visibility of gratings. *Journal of Physiology*, *197*, 551–566.

Canny, J. (1986). A computational approach to edge detection. *IEEE Transactions on Pattern Analysis and Machine Intelligence, PAMI-8*, 679–698.

Carpenter, G. A., and Grossberg, S. (eds.). (1991). *Pattern Recognition by Self-Organizing Neural Networks*. Cambridge, MA: MIT Press.

Cavanagh, P. (1987). Reconstructing the third dimension: Interactions between color, texture, motion, binocular disparity and shape. *Computer Vision, Graphics, and Image Processing, 37*, 171–195.

Cavanagh, P., and Favreau, O. E. (1985). Perception of motion in equiluminous kinematograms. *Perception, 14*, 151–162.

Cavanagh, P., Tyler, C. W., and Favreau, O. E. (1984). Perceived velocity of moving chromatic gratings. *Journal of the Optical Society of America (a), 1*, 893–899.

Chiyokura, H., and Kimura, F. (1983). Design of solids with free-form surfaces. *Computer Graphics, 17*, 289–298.

Ciarlet, P. G. (1978). *The Finite Element Method for Elliptic Problems*. Amsterdam: North-Holland.

Clark, J. J., and Yuille, A. L. (1990). *Data Fusion for Sensory Information Processing Systems*. Boston: Kluwer.

Clearwater, S. H., Huberman, B. A., and Hogg, T. (1991). Cooperative solution of constraint satisfaction problems. *Science, 254*, 1181–1183.

Clough, R. W., and Tocher, J. L. (1965). *Finite element stiffness matrices for the analysis of plate bending*. Paper presented at the Proceedings of the 1st Conference on Matrix Methods in Structural Mechanics. pp. 515–545.

Cooper, L. A. (1975). Mental transformation of random two-dimensional shapes. *Cognitive Psychology, 7*, 20–43.

Cooper, L. A., and Shepard, R. N. (1973). The time required to prepare for a rotated stimulus. *Memory and Cognition, 1*, 246–250.

Courant, R., and Hilbert, D. (1953). Methods of mathematical physics. *London: Interscience*.

Cummins, R. (1983). *The Nature of Psychological Explanation*. Cambridge, MA: MIT Press.

Dasarathy, B. V. (1994). *Decision Fusion*. Los Alamitos, CA: IEEE Computer Society Press.

Davis, L. S. (1975). A survey of edge detection techniques. *Computer Graphics and Image Processing, 4*(3), 248–270.

Dayanand, S., Uttal, W. R., Shepherd, T., and Lunskis, C. (1994). A particle system model for combining edge information from multiple segmentation modules. *CVGIP: Graphical Models and Image Processing, 56*, 219–230.

de Boor, C. (1987). B-form basics. In G. Farin (ed.), *Geometric Modeling: Algorithms and New Trends* (pp. 131–148). Philadelphia: SIAM.

Deutsch, E. S. (1972). Thinning algorithms on rectangular, hexagonal, and triangular arrays. *Communications of the ACM, 15*(9), 827–837.

De Valois, R. L. and DeValois, K. K. (1988). *Spatial Vision*. New York: Oxford University Press.

De Valois, R. L., and DeValois, K. K. (1993). A multi-stage color model. *Vision Research, 33*, 1053–1066.

De Valois, R. L., and Jacobs, G. H. (1984). Neural mechanisms of color vision. In J. M.

Brookhart & V. B. Mountcastle (eds.), *Handbook of Physiology—The Nervous System III* (pp. 425–456). Bethesda, MD: American Physiological Society.

De Valois, R. L., Jacobs, G. H., and Abramov, I. (1964). Responses of single cells in the visual system to shifts in the wavelength of light. *Science, 146*, 1184–1186.

De Valois, R. L., Abramov, I., and Jacobs, G. H. (1966). Analysis of response patterns of LGN cells. *Journal of the Optical Society of America, 56*, 966–977.

Dillon, C., and Caelli, T. (1992). *Generating complete depth maps in passive vision systems*. Paper presented at the Proceedings of the 11th IAPR International Conference on Pattern Recognition, The Hague, Netherlands, Aug. 30–Sept. 3.

Dretske, F. (1988). *Explaining Behavior: Reasons in a World of Causes*. Cambridge, MA: MIT Press.

Duchon, J. (1975). Splines minimizing rotation invariant semi-norms in Sobolev spaces. In W. Schempp and K. Zeller (eds.), *Multivariate Approximation Theory*. Basel, Switzerland, Birkhauser.

Durbin, R., and Willshaw, D. (1987). An analogue approach to the traveling salesman problem using an elastic net method. *Nature, 327*, 689–691.

Dyn, N., Levin, D., and Rippa, S. (1990). Algorithms for the construction of data-dependent triangulations. In *Algorithms for Approximation II* Mason J. C. and Cox, M. E. (eds.), (pp. 185–192). London: Chapman and Hall.

Estes, W. K., Campbell, J. A., and Hatsopoulos, N. (1989). Base-rate effects in category learning: A comparison of parallel network and memory storage-retrieval models. *Journal of Experimental Psychology: Learning, Memory, and Cognition, 15*, 556–571.

Farin, G. (1985). A modified Clough–Tocher interpolant. *Computer-Aided Geometric Design, 2*, 19–27.

Farin, G. (1990). Surfaces over Dirichlet tesselations. *Computer-Aided Geometric Design, 7*, 281–292.

Farin, G. (1993). *Curves and Surfaces in Computer Aided Geometric Design, Third Ed.* Boston: Academic Press.

Farin, G., and Kashyap, P. (1992). An iterative Clough–Tocher interpolant. *Mathematical Modeling and Numerical Design, 26*(1), 201–209.

Farley, B. G., and Clark, W. A. (1954). Simulation of a self-organizing system by a digital computer. *Institute of Radio Engineers Transactions of Information Theory, 4*, 76–84.

Felleman, D. J., and Van Essen. (1991). Distributed hierarchical processing in primate visual cortex. *Cerebral Cortex, 1*, 1–47.

Fendrich, R., Wessinger, C. M., and Gazzaniga, M. S. (1992). Residual vision in a scotoma: Implications for blindsight. *Science, 258*, 1489–1491.

Fischler, M. A., and Bolles, R. C. (1986). Perceptual organization and curve partitioning. *IEEE Transactions on Pattern Analysis and Machine Intelligence, 8*, 100–105.

Fitzpatrick, J. M., and Louze, M. R. (1987). A class of one-to-one two-dimensional transformations. *Computer Vision, Graphics, and Image Processing, 39*, 369–382.

Foley, J. D., van Dam, A., Feiner, S. K., and Hughes, J. F. (1990). *Computer Graphics, Principles and Practice* (2nd ed.). Boston: Kluwer Academic.

Foley, T. A., and Opitz, K. (1992). Hybrid cubic Bezier triangle patches. In T. Lyche and L. L. Schumaker (eds.), *Mathematical Methods in CAGD and Image Processing II*. Boston: Academic Press.

Fox, C. (1987). *An introduction to the calculus of variations*. New York: Dover.

Fram, J. R., and Deutsch, E. S. (1975). On the qualitative evaluation of edge detection schemes and their comparison with human performance. *IEEE Trans. Computing, C-24*(6), 616–628.

Franke, R. H. (1982a). Scattered data interpolation: Tests of some methods. *Math. Comp., 38*, 181–200.

Franke, R. H. (1982b). Smooth interpolation of scattered data by local thin plate splines. *Computer. Mathematics with Applications, 8*(4), 273–281.

Franke, R. H. (1987). Recent advances in the approximation of surfaces from scattered data. In *Topics in Multivariate Approximation* Chui, C. K., Shumaker, L. L. and Utreras, F. J. (eds.) (pp. 175–184). New York: Academic Press.

Franke, R. H., and Nielson, G. M. (1991). Scattered data interpolation: A tutorial and survey. In H. Hagen and D. Roller (eds.), *Topics in Multivariate Approximation* (pp. 131–160). New York: Springer.

Freeman, H. (1960). *Techniques for the digital computer analysis of chain-encoded arbitrary plane curves*. Paper presented at the Proceedings of the National Electronics Conference, Chicago, IL.

Freeman, W. J. (1975). *Mass Action in the Nervous System*. New York: Academic Press.

Freeman, W. J. (1995). *Societies of Brains: A Study in the Neuroscience of Love and Hate*. Hillsdale, NJ: Erlbaum.

Gamble, E. B., Geiger, D., Poggio, T., and Weinshall, D. (1989). Integration of vision modules and labeling of surface discontinuities. *IEEE Transactions on Systems, Man, and Cybernetics, 19*, 1576–1581.

Gardner, W. (1988). *Statistical spectral analysis: A nonprobabilistic theory*. Englewood Cliffs, NJ: Prentice-Hall.

Gegenfurter, K. R., and Hawken, M. J. (1995). Temporal and chromatic properties of motion mechanisms. *Vision Research, 35*, 1547–1563.

Geman, S., and Geman, D. (1984). Stochastic relaxation, Gibbs distributions, and the Bayesian restoration of images. *IEEE Transactions on pattern analysis and machine intelligence, 6*, 721–742.

Geyer, L. H., and DeWald, C. G. (1973). Subcomponents lists and confusion matrices. *Perception and Psychophysics, 14*, 471–482.

Gingerich, O. (1996). Wonders: Neptune, Velikovsky, and the name of the game. *Scientific American, September*, 181–183.

Goldmeier, E. (1972). Similarity of visually perceived forms. *Psychological Issues, 8*. (Whole #29).

Golub, G. H., and van Loan, C. (1983). *Matrix Computations*. (2nd ed.). Oxford: North Oxford Academic.

Gordon, J., and Shortliffe, E. H. (1990). The Dempster–Shafer theory of evidence. In G. Shafer and J. Pearl (eds.), *Readings in Uncertain Reasoning*. San Mateo, CA: Morgan-Kauffman.

Graham, N. (1980). Spatial-frequency channels in human vision: Detecting edges without edge detectors. In C. S. Harris (ed.), *Visual Coding and Adaptability* (pp. 215–262). Hillsdale, NJ: Erlbaum.

Graham, N. (1989). *Visual Pattern Analyzers*. New York: Oxford University Press.

Grandine, T. A. (1987). An iterative method for computing multivariate C^1 piecewise polynomial interpolants. *Computer Aided Geometric Design, 4*, 307–319.

Gray, S. B. (1971). Local properties of binary images in two dimensions. *IEEE Transactions on Computers, 20*(5), 551–561.

Grenander, U. (1993). *General Pattern Theory: A Mathematical Study of Regular Structures*. New York: Oxford University Press.

Grimson, W. E. L. (1981). *From Images to Surfaces: A Computational Study of the Human Early Visual System*. Cambridge, MA: MIT Press.

Grossberg, S. (1968). Some nonlinear networks capable of learning a spatial pattern of arbitrary complexity. *Proceedings of the National Academy of Sciences, 59*, 368–372.

Grossberg, S. (1969). On learning of spatiotemporal patterns by networks with ordered sensory and motor components. *Studies in Applied Mathematics, 49*, 135–166.

Grossberg, S. (1982). *Studies of Mind and Brain: Neural Principles of Learning, Perception, Development, Cognition, and Motor Control*. Boston: Reidel Press.

Grossberg, S. (1988a). Nonlinear neural networks: Principles, mechanisms, and architectures. *Neural Networks, 1*, 17–61.

Grossberg, S. (1988b). *Neural Networks and Natural Intelligence*. Cambridge, MA: MIT Press.

Grossberg, S., and Mingolla, E. (1993). Neural dynamics of motion perception: Direction fields, apertures, and resonant grouping. *Perception and Psychophysics, 53*, 243–278.

Gudder, S. (1988). *Quantum probability*. New York: Academic Press.

Hagen, H., Schreiber, T., and Geschwind, E. (1990). Methods for surface interrogation. In A. Kaufman (ed.), *Visualization 90* (pp. 187–193). Los Alamitos, CA: IEEE Press.

Haralick, R. M. (1980). Edge and region analysis for digital image data. *Computer Graphics and Image Processing, 12*(1), 60–73.

Hardy, R. L. (1970). Multiquadric equations of topography and other irregular surfaces. *Journal of Geophysics Research, 76*, 1905–1915.

Hardy, R. L. (1990). Theory and applications of the multiquadric-biharmonic methods. *Computer Mathematics Applications, 19*, 163–208.

Hartline, H. K. (1949). Inhibition of activity of visual receptors by illuminating nearby retinal areas in the Limulus eye. *Federation Processing, 8*, 69.

Hashimoto, M., and Sklansky, J. (1987). Multiple-order derivatives for detecting local image characteristics. *Computer Vision, Graphics, and Image Processing, 39*, 28–55.

Haugeland, J. E. (1981). *Mind Design: Philosophy, Psychology, Artificial Intelligence*. Cambridge, MA: MIT Press.

Hecht, S. (1931). Die physikalische Cemie und die physiologie des sehaktes. *Erg. Physiol., 32*, 243–290.

Hecht, S., Shalaer, S., and Pirenne, M. H. (1942). Energy, quanta, and vision. *Journal of General Psychology, 25*, 819–840.

Helmholtz, H. v. (1867;1925). *Excerpts from Treatises on Physiological Optics* (3rd ed.). New York: Wiley.

Henderson, M. R., and Anderson, D. C. (1984). Computer recognition and extraction of form subcomponents: A CAD/CAM link. *Computers in Industry, 5*, 329–339.

Hering, E. (1878:1964). *Outline of a Theory of the Light Sense*. (Translated by L. M. Hurvich, and D. Jameson). Cambridge, MA.: Harvard University Press.

Hilgetag, C.-C., O'Neill, M. A., and Young, M. P. (1996). Indeterminate organization of the visual system. *Science, 271*, 776–777.

Hochberg, J. E., and McAlister, E. (1953). A quantitative approach to figural "goodness." *Journal of Experimental Psychology, 46*, 361–364.

Hock, H. S., Webb, E., and Cavedo, L. C. (1987). Perceptual learning in visual category acquisition. *Memory and Cognition, 15*, 544–556.

Hoffman, D. D., and Richards, W. A. (1984). Parts of recognition. *Cognition, 18*, 65–96.

Hoffman, W. C. (1966). The Lie algebra of visual perception. *Journal of Mathematical Psychology, 3*, 349–367.

Hoffman, W. C. (1980). Subjective geometry and geometric psychology. *Mathematical Modeling, 1*, 349–367.

Homa, D., Dunbar, S., and Nohre, L. (1991). Instance frequency, categorization, and the modulating effect of experience. *Journal of Experimental Psychology: Learning, Memory, and Cognition, 17*, 444–458.

Horgan, J. (1996). *The End of Science*. Reading, MA: Addison-Wesley.

Horn, B. K. P. (1970). *Shape from Shading: A Method for Obtaining the Shape of a Smooth Opaque Object from One View*. Unpublished Ph.D. Dissertation, MIT, Cambridge, MA.

Horn, B. K. P. (1990). Height and gradient from shading. *International Journal of Computer Vision, 5*, 584–595.

Horn, B. K. P., and Brooks, M. J. (1989). *Shape from Shading*. Cambridge, MA: MIT Press.

Huntsberger, T. L., Jacobs, C. L., and Cannon, R. L. (1985). Iterative fuzzy image segmentation. *Pattern Recognition, 18*(2), 131–138.

Hurvich, L. M., and Jameson, D. (1955). Some quantitative aspects of an opponent color theory. II. Brightness, saturation, and hue in normal and dichromatic vision. *Journal of the Optical Society of America, 45*, 602–616.

Jameson, D. and Hurvich, L. M. (1955). Some quantitative aspects of an opponent-colors theory. *Journal of the Optical Society of America, 45*, 546–552.

Jeong, H., and Kim, J.-G. (1992). *A unification theory for early vision*. Paper presented at the Proceedings of the 1992 IEEE/RSJ International Conference on Intelligent Robots and Systems, July, 1992, Raleigh, N. C.

Judd, D. B. (1951). Basic correlates of the visual stimulus. In S. S. Stevens (ed.), *Handbook of Experimental Psychology* (pp. 811–867). New York: Wiley.

Julesz, B. (1971). *Foundations of Cyclopean Perception*. Chicago: University of Chicago Press.

Julesz, B. (1981). Textons, the elements of texture perception, and their interactions. *Nature, 290*, 91–97.

Kaas, J. H. (1978). *The organization of visual cortex in primates*. New York: Plenum Press.

Kaas, J. H., and Heurta, M. F. (1988). *Comparative Primate Biology, 4*, 327.

Kabrisky, M. (1966). *A proposed Model for Information Processing in the Human Brain*. Urbana: University of Illinois Press.

Kanizsa, G. (1979). *Organization of Vision. Essays on Gestalt Perception*. New York: Praeger.

Kass, M., Witkin, A., and Terzopoulos, D. (1987). Snakes: Active edge models. *International Journal of Computers, 1*, 321–331.

Kirkpatrick, S., Gellatt, C. D., and Vecchi, M. P. (1982). *Optimization by simulated annealing*. Yorktown Hts., N.Y.: IBM Thomas J. Watson Research Center.

Koffka, K. (1935). *Principles of Gestalt Psychology*. New York: Harcourt, Brace, and World.

Kolmogorov, A. N. (1929:1956). *Foundations of the Theory of Probability*. New York: Chelsea.

Kong, T. Y., and Rosenfeld, A. (1989). Digital topology: Introduction and survey. *Computer Vision, Graphics and Image Processing, 48*, 357–393.

Krumhansl, C. L. (1978). Concerning the applicability of geometric models to similarity data: The interrelationship between similarity and spatial density. *Psychological Review, 85*, 445–463.

Kubovy, M., and Pomerantz, J. R. (1981). *Perceptual Organization*. Hillsdale, NJ: Erlbaum.

Land, E. H. (1977). The Retinex theory of color vision. *Scientific American, 237*, 108–128.

Landy, M. S., and Movshon, J. A. (eds.). (1991). *Computational Models of Visual Processing*. Cambridge, MA: MIT Press.

Lawson, C. L. (1977). *Software for C1 surface interpolation*. New York: Academic Press.

Lawson, C. L. (1986). Properties of *n*-dimensional triangulations. *Computer-Aided Geometric Design, 3*, 231–246.

Lee, C. H., and Rosenfeld, A. (1989). Improved methods of estimating shape from shading using the light source coordinate system. In B. K. P. Horn and M. J. Brooks (eds.), *Shape from Shading* (pp. 323–348). Cambridge, MA: MIT Press.

Li, X., and Zhu, Z. (1988). Group direction difference chain codes for the representation of the border. *Proceedings of SPIE, 938*, 372–376.

Link, S. W. (1992). *The Wave Theory of Difference and Similarity*. Hillsdale, NJ: Erlbaum.

Livingstone, M., and Hubel, D. (1987). Connections between layer 4b of area 17 and the thick cytochrome oxidase stripes of area 18 in the squirrel monkey. *Journal of Neuroscience, 7*, 3371–3377.

Livingstone, M., and Hubel, D. (1988). Segregation of form, color, movement, and depth: Anatomy, physiology, and perception. *Science, 240*, 740–749.

Lovell, R., Uttal, W. R., Shepherd, T., and Dayanand, S. (1992). A model of texture discrimination using multiple weak operators and spatial averaging. *Pattern Recognition, 25*, 1157–1170.

Magoun, H. W. (1954). The ascending reticular system and wakefulness. In J. F. Delafresnaye (ed.), *Brain Mechanisms and Consciousness*. Springfield, IL: Charles C. Thomas.

Malt, B. C. (1989). An on-line investigation of prototype and exemplar strategies in classification. *Journal of Experimental Psychology: Learning, Memory, and Cognition, 15*, 539–555.

Mäntylä, M. (1988). *An Introduction to solid Modeling*. Rockville, MD, Computer Science Press.

Maragos, P. (1988). Optimal morphological approaches to image matching and object detection. *International Conference on Computer Vision*, 695–699.

Marks, W. B., Dobelle, W. H., and MacNichol, E. F. (1964). Visual pigments of single primate cones. *Science*, *143*, 1181–1183.

Marr, D. (1982). *Vision*. San Francisco: Freeman.

Marr, D., and Hildredth, E. C. (1980). *Theory of edge detection*. Paper presented at the Proceedings of the Royal Society of London B.

Marr, D., and Nishihara, H. K. (1978). *Representation and recognition of the spatial organization of three-dimensional shapes*. Paper presented at the Proceedings of the Royal Society of London B.

Marr, D., and Poggio, T. (1979). *A computational theory of human Stereo vision*. Paper presented at the Proceedings of the Royal Society of London B.

Matthies, L., Szelsiki, R., and Kanade, T. (1988). *Incremental estimation of dense depth maps*. Paper presented at the IEEE Conference on Computer Vision and Pattern Recognition,.

McClelland, J. L., Rumelhart, D. E., and PDP Group, (1986). *Parallel Distributed Processing. Volume 2: Psychological and Biological Models*. Cambridge, MA: MIT Press.

Mendel, J. M. (1995). Fuzzy logic systems for engineering. *Proceedings of the IEEE*, 84, 345–377.

Metropolis, N., Rosenbluth, A. W., Rosenbluth, M. N., Teller, A. H., and Teller, E. (1953). Equations of state calculations by fast computing machines. *Journal of Chem. Phys.*, *21*, 1087–1091.

Metzler, J., and Shepard, R. N. (1974). Transformational studies of the internal representations of three dimensional objects. In R. L. Solso (ed.), *Theories of Cognitive Psychology: The Loyola Symposium*. Hillsdale, NJ: Erlbaum.

Moore, E. F. (1956). Gedanken-experiments on sequential machines. In C. E. Shannon and J. McCarthy (eds.), *Automata Studies* (pp. 129–153). Princeton, NJ: Princeton University Press.

Morrone, M. C., and Burr, D. C. (1988). Subcomponent detection in human vision: A phase-dependent energy model. *Proceedings of the Royal Society of London, Series B*, *235*, 221–245.

Moruzzi, G. (1954). The physiological properties of the brain stem reticular system. In J. F. Delafresnaye (ed.), *Brain Mechanisms and Consciousness*. Springfield, IL: Charles C. Thomas.

Mueller, J. (1840). *Handbuch der physiologie des Menschen* (*Vol. II*). Coblentz, Germany: Holscher.

Murphy, G. L., and Wisniewski, E. J. (1989). Categorizing objects in isolation and in scenes: What a superordinate is good for. *Journal of Experimental Psychology: Learning, Memory, and Cognition*, *15*, 572–586.

Nalwa, V. S., and Pauchon, E. (1987). Edge aggregation and edge description. *Computer Vision, Graphics, and Image Processing*, *40*, 79–94.

Neitz, M., Neitz, J. and Jacobs, G. H. (1991). Spectral tuning of pigments underlying red-green color vision. *Science*, *252*, 971–974.

Newell, A. (1990). *Unified Theories of Cognition*. Cambridge, MA: Harvard University Press.

Newhouse, M., and Uttal, W. R. (1982). Distribution of stereoanomalies in the general population. *Bulletin of the Psychonomic Society*, *20*, 48–50.

Niall, K. K. (1998). "Mental rotation," pictured rotation, and tandem rotation in depth. (personal communication).

Nielson, G. M. (1979). The side-vertex method for interpolation in triangles. *Journal of Approximation Theory*, *25*, 318–336.

Nielson, G. M. (1980). Minimum norm interpolation in triangles. *SIAM Journal of Numerical Analysis*, *17*, 46–62.

Nielson, G. M. (1983). A method for interpolating scattered data based upon a minimum norm network. *Mathematics of Computation*, *40*, 253–271.

Nielson, G. M. (1993). Scattered data modeling. *IEEE Computer Graphics and Applications*, *13*(1), 60–70.

Nielson, G. M., and Ramaraj, R. (1987). Interpolation over a sphere based upon a minimum norm network. *Computer-Aided Geometric Design*, *4*, 41–57.

Nielson, G. M., Foley, T. A., Hamann, B., and Lane, D. (1991). Visualizing and modeling of scattered multivariate data. *IEEE Computer Graphics and Applications*, *11*(3), 47–55.

Nosofsky, R. M. (1985). Overall similarity and the identification of separable dimension stimuli: A choice model analysis. *Perception & Psychophysics*, *38*, 415–432.

Nosofsky, R. M. (1988). Exemplar-based accounts of relations between classification, recognition, and typicality. *Journal of Experimental Psychology: Learning, Memory, and Cognition*, *14*, 700–708.

Nosofsky, R. M., Clark, S. E., and Shin, H. J. (1989). Rules and exemplars in categorization, identification, and recognition. *Jrnl. of Exp. Psych: Learning, Memory and Cognition*, *15*, 282–304.

Optical Society of America, Committee on Colorimetry (1953). *The Science of Color*. New York: Crowell.

Penfield, W., and Roberts, L. (1959). *Speech and Brain Mechanisms*. Princeton, NJ: Princeton University Press.

Pentland, A. P. (1984). Local shading and analysis. *IEEE Trans. on Pattern Analysis and Machine Intelligence*, *6*, 170–187.

Poggio, T., and Girosi, F. (1990). Networks for approximation and learning. *Proceedings of the IEEE*, *78*, 1481–1497.

Poggio, T., Torre, V., and Koch, C. (1985). Computational vision and regularization theory. *Nature*, *317*, 214–319.

Poggio, T., Gamble, E. B., and Little, J. J. (1988). Parallel integration of vision modules. *Science*, *242*, 436–440.

Polyak, S. L. (1941). *The Retina*. Chicago: University of Chicago Press.

Polyak, S. L. (1957). *The Vertebrate Visual System*. Chicago: University of Chicago Press.

Pottmann, H. (1991). Interpolation of scattered data based on generalized minimum norm networks. *Constr. Approx.*, 247–256.

Pottmann, H. (1992). Interpolation on surfaces using minimum norm networks. *Computer-Aided Geometric Design*, *2*, 51–67.

Prenter, P. M. (1985). *Splines and Variational Methods*. New York: Wiley.

Prewitt, J. M. S. (1970). Object enhancement and extraction. In B. S. Lipkin and A.

Rosenfeld (eds.), *Picture Processing and Psychopictorics* (pp. 75–150). New York: Academic Press.

Price, K. E. (1985). Relaxation matching techniques: A comparison. *IEEE Transactions on Pattern Analysis and Machine Intelligence*, *7*, 617–623.

Quak, E., and Schumaker, L. L. (1989). Calculation of the energy of a piecewise polynomial surface. In *Algorithms for Approximation II* M. G. Cox, Mason, J. C. (eds.), (pp. 134–143). New York: Clarendon Press.

Quak, E., and Schumaker, L. L. (1990). Cubic spline fitting using data-dependent triangulations. *Computer-Aided Geometric Design*, *7*, 293–301.

Ramachandran, V. S. (1985). The neurobiology of perception. *Perception*, *14*, 97–103.

Ramachandran, V. S. (1988). Perception of shape from shading. *Science*, *331*, 163–166.

Ratoosh, P. (1949). On interposition as a cue for the perception of distance. *Proceedings of the National Academy of Science, Washington*, *35*, 257–259.

Rearick, T. C., Frawley, J. L., and Cortopassi, P. P. (1988). Using perceptual grouping to recognize and locate partially occluded objects. *IEEE Conf. on Computer Vision and Pattern Recognition* 840–846.

Reeves, W. T. (1983). Particle systems—A technique for modeling a class of fuzzy objects. *Siggraph*, *83*, 359–376.

Renka, R. L., and Cline, A. K. (1984). A triangle-based C1 interpolation method. *Rocky Mountain Journal of Mathematics*, *14*(1), 223–238.

Richards, W. (1984). Structure from stereo and motion. *Journal of the Optical Society of America*, *3*, 343–349.

Roberts, L. G. (1965). Machine perception of three-dimensional solids. In J. T. Tippett (ed.), *Optical and Electro-optical Information Processing*. Cambridge, MA: MIT Press.

Rodieck, R. W. (1973). *The Vertebrate Retina: Principles of Structure and Function.* San Francisco: Freeman.

Rosch, R., and Mervis, C. B. (1975). Family resemblances: Studies in the internal structure of categories. *Cognitive Psychology*, *7*, 573–603.

Rosenblatt, F. (1962). *The Principles of Neurodynamics*. Washington, DC: Spartan Books.

Rosenfeld, A., and Kak, A. (1982). *Digital Picture Processing*. New York: Academic Press.

Ruch, T. C. and Patton, H. D. (1965). *Physiology and Biophysics*. Philadelphia: W. B. Saunders.

Rumelhart, D. E., McClelland, J. L., and PDP Group. (1986). *Parallel Distributed Processing. Volume 1: Foundations*. Cambridge, MA: MIT Press.

Rushton, W. A. H. (1972). Visual pigments in man. In H. J. A. Dartnall (ed.), *Photochemistry of Vision* (pp. 364–394). Berlin: Springer Verlag.

Safranek, R. J., Gottschalk, S., and Kak, A. C. (1990). Evidence accumulation using binary frames of discernment for verification vision. *IEEE Transactions on robotics and automation*, *6*(4), 405–417.

Sahoo, P. K., Soltani, S., Wong, A. K. C., and Chen, Y. C. (1988). A survey of thresholding techniques. *Computer Vision, Graphics, and Image Processing*, *41*, 233–260.

Sakitt, B. (1972). Counting every quantum. *Journal of Physiology*, *223*, 131–150.

Salzman, M. (1912). *The Anatomy and Physiology of the Human Eyeball in the Normal State*. Chicago: University of Chicago Press.

Schonemann, P. H., Dorcey, T., and Kienapple, K. (1985). Subadditive concatenation in dissimilarity judgments. *Perception and Psychophysics, 38*, 1–17.

Schroedinger, E. (1925). Weber das Verhaltnis der Vierfarben-zur Dreifarbentheorie. *Sitzber. Akad. Wiss. Wien, 134.*

Schumaker, L. L. (1976). Fitting surfaces to scattered data. In G. Lorentz, C. Chui, and L. L. Schumaker (Eds.), *Approximation Theory II* (pp. 203–268). New York: Academic Press.

Schumaker, L. L. (1993). Computing optimal triangulations using simulated annealing. *Computer-Aided Geometric Design, 10*, 329–345.

Sekuler, R., and Nash, D. (1972). Speed of size scaling in human vision. *Psychonomic Science, 27*, 93–94.

Selfridge, O. G. (1958). *Pandemonium: A paradigm for learning.* Paper presented at the Mechanization of Thought Processes Conference, London, (November, 1958).

Shafer, G. (1976). *A mathematical theory of evidence.* Princeton, NJ: Princeton University Press.

Shafer, G., and Pearl, J. (1990). *Readings in Uncertain Reasoning.* San Mateo, CA: Morgan Kaufman.

Shepard, R. N. (1957). Stimulus and response generalization: A stochastic model relating generalization to distance in a psychological space. *Psychometrika, 22*, 325–345.

Shepard, R. N., and Metzler, J. (1971). Mental rotation of three-dimensional objects. *Science, 171*, 701–703.

Shepherd, T., Uttal, W. R., Dayanand, S., and Lovell, R. (1992). A method for shift, rotation, and scale invariant pattern recognition using the form and arrangement of pattern-specific features. *Pattern Recognition, 25*, 343–356.

Sibson, R. (1981). A brief description of the natural neighbor interpolant. In V. Barnett (ed.), *Interpolating multivariate data.* New York: Wiley.

Skarda, C. A., and Freeman, W. J. (1987). How brains make chaos in order to make sense of the world. *Behavioral and Brain Sciences, 10*, 161–195.

Sobel, I. (1970). *Camera Models and Machine Perception.* Palo Alto, CA: Stanford University Press.

Stein, B. E., and Meredith, M. A. (1993). *The Merging of the Senses.* Cambridge, MA: MIT Press.

Stoner, G. R., and Albright, T. D. (1993). Image segmentation cues in motion processing: Implications for modularity. *Journal of Cognitive Neuroscience, 5*, 129–149.

Strang, G., and Fix, G. (1973). *An Analysis of the Finite Element Method.* Englewood Cliffs, NJ: Prentice-Hall.

Stromer, R., and Stromer, J. B. (1990a). The formation of arbitrary stimulus classes in matching to complex samples. *The Psychological Record, 40*, 51–66.

Stromer, R., and Stromer, J. B. (1990b). Matching to complex samples: Further study of arbitrary stimulus classes. *The Psychological Record, 40*, 505–516.

Tolhurst, D. J. (1972). On the possible existence of edge detector neurons in the human visual system. *Vision Research, 12*, 797–804.

Torgenson, W. S. (1965). Multidimensional scaling of similarity. *Psychometrika, 30*, 379–393.

Townsend, J. T. (1990). Chaos theory: A brief tutorial and discussion. In A. F. Healy,

S. M. Kosslyn, and R. M. Shiffrin (eds.), *From Learning Processes to Cognitive Process: Essays in Honor of W. K. Estes* (Vol. 1, pp. 65–96). Hillsdale, NJ: Erlbaum.

Treisman, A. (1986a). Features and objects in visual processing. *Scientific American, 255*, 114–125.

Treisman, A. (1986b). Properties, parts, and objects. In K. R. Boff, L. Kaufman, and J. P. Thomas (eds.), *Handbook of Perception and Human Performance* (pp. 1–70). New York: Wiley.

Treisman, A., and Gelade, G. (1980). A feature-integration theory of attention. *Cognitive Psychology, 12*, 97–136.

Treisman, A., and Patterson, R. (1984). Emergent features, attention, object perception. *Journal of Experimental Psychology: Human Perception and Performance, 10*, 12–31.

Tversky, A. (1977). Subcomponents of similarity. *Psychological Review, 84*, 327–352.

Ullman, S. (1979). *The interpretation of visual motion*. Cambridge, MA: MIT Press.

Uttal, W. R. (1973). *The Psychobiology of Sensory Coding*. New York: Harper & Row.

Uttal, W. R. (1978). *The Psychobiology of Mind*. Hillsdale, NJ: Erlbaum.

Uttal, W. R. (1981). *A Taxonomy of Visual Processes*. Hillsdale, NJ: Erlbaum.

Uttal, W. R. (1988). *On Seeing Forms*. Hillsdale, NJ: Erlbaum.

Uttal, W. R. (1998). *Toward a New Behaviorism: The Case Against Perceptual Reductionism*. Mahwah, NJ: Erlbaum.

Uttal, W. R., Davis, N. S., Welke, C., and Kakarala, R. (1988). The reconstruction of static visual forms from sparse dotted samples. *Perception and Psychophysics, 43*, 223–240.

Uttal, W. R., Bradshaw, G., Dayanand, S., Lovell, R., Shepherd, T., Kakarala, R., Skifsted, K., and Tupper, G. (1992). *The Swimmer: An Integrated Computational Model of a Perceptual-Motor System*. Hillsdale, NJ: Erlbaum.

Uttal, W. R., Liu, N., and Kalki, J. (1996). An integrated computational model of three-dimensional vision. *Spatial Vision, 9*, 393–422.

Van Essen, D. C. (1985). Functional organization of primate visual cortex. *Cerebral Cortex, 3*, 259–329.

Van Essen, D. C., Anderson, C. H., and Felleman, D. J. (1992). Information processing in the primate visual system: An integrated systems perspective. *Science, 255*, 419–423.

Voronoi, G. (1908). Nouvelle applications des parameters continues a la theorie des formes quadratiques. *Journal Reine Agnew. Math., 134*, 198–287.

Walls, G. L. (1942). *The Vertebrate Eye*. Bloomfield Hills, MI: Cranbrook Institute of Science.

Wandell, B. A. (1995). *Foundations of Vision*. Sunderland, MA: Sinauer Associates.

Wang, Y. F., Mitiche, A., and Aggarwal, J. K. (1987). Computation of surface orientation and structure of objects using grid coding. *IEEE Transactions On Pattern Analysis and Machine Intelligence, 2*, 129–137.

Watanabe, T. (1998). *High level motion processing*. Cambridge, MA. MIT Press.

Watson, A. B. (ed.). (1993). *Digital Images and Human Vision*. Cambridge, MA: MIT Press.

Weiskrantz, L., Warrington, E. K., Sander, M. D., and Marshall, J. (1974). Visual capacity in the hemianopic field following a restricted occipital ablation. *Brain*, *97*, 709–728.

Wertheimer, M. (1938). Laws of organization in perceptual forms. In W. D. Ellis (ed.), *A Sourcebook of Gestalt Psychology* (pp. 71–88). New York: Harcourt, Brace.

Widrow, B. (1962). *Generalization and Information Storage in Networks of Adaline Neurons*. Washington, DC: Spartan Books.

Wilson, H. R., and Gelb, D. J. (1984). Modified line element theory for spatial frequency and width discrimination. *Journal of the Optical Society of America (A)*, *1*, 124–131.

Wilson, H. R., Levi, D., Maffei, L., Rovamo, J., and DeValois, R. (1990). The perception of form: Retina to striate cortex. In Spillman, L. & Werner, J. J. (eds.), *Visual Perception: The Neurophysiological Foundations* (pp. 231–272). New York: Academic Press.

Wojcik, Z. M. (1984). An approach to the recognition of contours and line-shaped objects. *Computer Vision, Graphics, and Image Processing*, *25*(2), 184–204.

Woodham, R. J. (1980). Using digital terrain data to model image information in remote sensing. *Image Processing for Missile Guidance*. *SPIE*, *238*.

Woolsey, C. N. (1952). Pattern of localization in sensory and motor areas of the cerebral cortex. *The Biology of Mental Health and Disease: The Twenty-seventh Annual Conference of the Milbank Fund* (pp. 193–205). New York: Hoeber.

Worsey, A., and Farin, G. (1987). An *n*-dimensional Clough–Tocher interpolant. *Constructive Approximation*, *3*, 99–110.

Wu, R., and Stark, H. (1985). Rotation-invariant pattern recognition using optimum subcomponent extraction. *Applied Optics*, *24*, 179–184.

Young, R. W. (1970). Visual cells. *Scientific American*, *223*, 80–91.

Zadeh, L. A. (1973). Outline of a new approach to the analysis of complex systems and decision processes. *IEEE Transactions on systems, man, and cybernetics*, *SMC-3*(1), 28–44.

Zeki, S. (1993). *A Vision of the Brain*. Oxford: Blackwell.

Zheng, Q. F., and Chellappa. (1991). Estimation of illuminant direction, albedo, and shape from shading. *IEEE Transactions on Pattern Analysis and Machine Intelligence*, *13*, 680–702.

Zucker, S. W., David, C., Dobbins, A., and Iverson, C. (1988). *The organization of curve detection: Coarse tangent fields and fine spline coverings*. Paper presented at the International Conference on Computer Vision, 568–577.

Index